Bioethical Issues in Healthcare

Edited by Peter A. Clark

Published in London, United Kingdom

IntechOpen

Supporting open minds since 2005

Bioethical Issues in Healthcare
http://dx.doi.org/10.5772/intechopen.95187
Edited by Peter A. Clark

Contributors
Akram Jassim Jawad, Carlo Ciardo, Oludayo Tade, Mary Jo Iozzio, Md. Shah Alam Panna, Una P. Canning, Federico D'Angiolillo, Mrinmoy Roy, Peter A. Clark, David Grana

First published in London, United Kingdom, 2022 by IntechOpen
IntechOpen is the global imprint of INTECHOPEN LIMITED, registered in England and Wales, registration number: 11086078, 5 Princes Gate Court, London, SW7 2QJ, United Kingdom
Printed in Croatia

British Library Cataloguing-in-Publication Data
A catalogue record for this book is available from the British Library

Additional hard and PDF copies can be obtained from orders@intechopen.com

Bioethical Issues in Healthcare
Edited by Peter A. Clark
p. cm.
Print ISBN 978-1-83969-466-0
Online ISBN 978-1-83969-467-7
eBook (PDF) ISBN 978-1-83969-468-4

For EU product safety concerns:
IN TECH d.o.o., Prolaz Marije Krucifikse Kozulić 3, 51000 Rijeka, Croatia, info@intechopen.com or visit our website at intechopen.com.

Meet the editor

Peter A. Clark, SJ, Ph.D., is the John McShain Chair in Ethics and Director of the Institute of Clinical Bioethics, Saint Joseph's University, Philadelphia. He is also the bioethicist for the Triunity Health System of Philadelphia, Jefferson Health System, St. Christopher's Hospital for Children and Shriners Hospital for Children in Philadelphia, St. Agnes Hospital in Baltimore, Mercy Hospital in Baltimore, Catholic Charites of Maryland, and Caritas Baby Hospital in Bethlehem, Palestine. As a bioethicist, Dr. Clark is responsible for the ethical training of the medical fellows, residents, and interns in all the acute care hospitals. He does weekly Ethics Teaching Rounds at the four acute care facilities in the Trinity Health System and the three acute care hospitals in the Jefferson Health System Northeast. He also co-chairs hospital ethics committees, IRBs, and the Corporate Ethics Committee and is on consult 24/7 for all the hospitals. Dr. Clark received his Ph.D. in Medical Ethics from the Loyola University of Chicago. He is the author of numerous articles in medical and bioethics journals on topics including medical futility, pain management, prejudice in the medical profession, the medical use of marijuana, safe injection sites, opioid prevention and education, tube feedings and PVS patients, male circumcision and HIV/AIDS, face transplantation, the Ashley Treatment, and more. He is also the author of *To Treat or Not to Treat: The Ethical Methodology of Richard A. McCormick as Applied to Treatment Decisions for Handicapped Newborns* and *Death With Dignity: Ethical and Practical Considerations for Caregivers of the Terminally Ill*. Dr. Clark is also the editor of Contemporary Issues in Bioethics and Bioethics: Medical, Ethical and Legal Perspectives. He is also a reviewer for numerous bioethics journals and is a member of various medical and ethical organizations nationally and internationally.

Contents

Preface XI

Section 1
End-of-Life Issues 1

Chapter 1 3
An Examination of Safe Injection Sites and Ethical Issues in Philadelphia, United States
by Peter A. Clark and David Grana

Chapter 2 31
Determination of Death: Ethical and Biomedical Update with International Consensus
by Md. Shah Alam Panna

Chapter 3 59
Compassion Versus Care in Healthcare Institutions: What's the Difference?
by Una P. Canning

Section 2
Inequalities and Disparities in Healthcare 77

Chapter 4 79
The Risk of Inequality in Italian Healthcare Due to Covid-19
by Carlo Ciardo

Chapter 5 89
Radical Bioethics: Difference, Disability, and Desiderata
by Mary Jo Iozzio

Section 3
Models in Healthcare 109

Chapter 6 111
For a Model of Revision, Assistance and Care of Identities
by Federico D'Angiolillo

Chapter 7 129
COVID-19 Myths and Music Advocacy in Nigeria
by Oludayo Tade

Section 4
New Technologies **139**

Chapter 8 **141**
Bioethics of Brain Computer Interfaces
by Akram Jassim Jawad

Chapter 9 **149**
Good Pharmacy Practice in India: Its Past, Present and Future with Need and Status in COVID 19
by Mrinmoy Roy

Preface

Bioethics is a relatively new discipline of ethics dealing with ethical problems arising from medicine and medical research. Healthcare professionals are being confronted daily with ethical dilemmas that arise from their relationships with patients, clinical institutions, the pharmaceutical industry, biotech firms, and so on. These ethical issues encompass medical futility, genetic engineering, stem cell research, organ transplantation, nanotechnology, allocation of scarce resources, inequalities in healthcare, beginning-of-life issues, and end-of-life issues. To analyze these issues, bioethicists use various ethical theories to assist them in finding clarity and consistency in their decision-making. These theories may differ in approach, but they all utilize established ethical rules and principles to assure prudent and coherent solutions in clinical situations where different people's interests or priorities conflict. These ethical principles include autonomy, beneficence, nonmaleficence, confidentiality, justice, and so on. These principles serve as the backbone for these ethical theories, and they assist us in applying a holistic approach and coming to well-reasoned positions.

The various chapters in this book present ethical dilemmas that healthcare professionals confront today and will confront in the future. Various perspectives are presented by the authors, which not only can assist healthcare professionals but also challenge them to explore and rethink some of the basic ideas in medicine and medical research and pave the way into new and uncharted territories. Technological and medical developments are happening at a very fast pace in bioethics and we are being called to process our progress at a warp speed. The speed of progress is sometimes outpacing our ability to ethically discern the rightness or wrongness of our actions. As a result, we are seeing the potential for depersonalization and dehumanization of people who are the most vulnerable in society. The central issue is that scientists are not going to stop the speed of progress. Therefore, bioethicists must answer the call to balance risks and benefits and the nature of the human person and determine where to draw lines between treatments that are therapeutic versus those that involve enhancement and how to distribute our limited medical resources in a just manner globally. To accomplish this task, dialogue and transparency must continue to be enhanced and encouraged among bioethicists, researchers, scientists, and clinicians. The issues confronting us now and in the future in medicine and research are only going to become more complex, comprehensive, and challenging. It is the role of the bioethicist today to engage in a constructive dialogue with their colleagues in medicine, scientific research, law, and the biomedical industrial complex so that new procedures and techniques are completely vetted from all vantage points to preserve and defend the very dignity and respect of the human person. This job may appear to be daunting at first glance, but unless those trained in philosophical and theological ethics take the lead, the future of humanity will become compromised. This book is an attempt to encourage a dialogue with our colleagues in various disciplines to examine these critical bioethical issues from all sides so that the best interest of patients, families, and society as a whole will be protected in the future.

Peter A. Clark, S.J., Ph.D.
Professor of Medical Ethics,
John McShain Chair in Ethics,
Director-Institute of Clinical Bioethics,
Saint Joseph's University,
Philadelphia, Pennsylvania

Section 1

End-of-Life Issues

Chapter 1

An Examination of Safe Injection Sites and Ethical Issues in Philadelphia, United States

Peter A. Clark and David Grana

Abstract

The opioid epidemic in the United States has been an ever-increasing public health crisis. Despite being a major issue in the United States for decades, relatively little action has been taken to address the opioid crisis. To mitigate the harm the opioid epidemic has caused in the United States, safe injection sites have emerged as a promising solution. Despite the exhaustive benefits of safe injection sites, including the reduction in the number of opioid overdose deaths, safe injection sites have faced opposition in the United States. Most of these concerns in the United States question the legality of safe injection sites, along with potential community implications. Through examining the ethics of safe injection sites from a Catholic social teaching perspective and performing an integrative literature review, safe injection sites are clearly ethical and would aid in respecting the dignity and life of people who inject drugs (PWID). With safe injection sites being ethical and recommendations in this paper to overcome concerns about safe injection sites, safe injection sites are a viable option to combat the opioid crisis in the United States.

Keywords: safe injection sites, opioid epidemic, opioids, opioid-related disorders, harm reduction

1. Introduction

1.1 Ethical principles from a catholic social teaching perspective

While public health crises, such as the COVID-19 pandemic, have received tremendous attention and resources, one public health crisis has seemingly been forgotten for decades: The opioid epidemic. With nearly half a million Americans dead from opioid use in the past few decades and no foreseeable stop to the increasing number of opioid overdose deaths, the United States is in need of immediate solutions to the opioid crisis. One emerging solution to combat the opioid crisis is facilities known as safe injection sites. Safe injection sites allow people who inject drugs (PWID) to safely inject addictive substances in the presence of healthcare professionals. While safe injection sites have been successful in countries across the world for decades as a solution to the opioid crisis, they have recently been a topic of debate in the United States. Especially in cities, such as Philadelphia, where the opioid crisis is one of the worst in the nation, safe injection sites are an extremely promising solution.

Before safe injection sites are able to be discussed in detail, this paper will first provide a background to the opioid epidemic, highlighting the role of pharmaceutical companies, physicians, the Drug Enforcement Agency (DEA), and the Food and Drug Administration (FDA) in exacerbating the opioid crisis in the United States. The three stages of the rise in opioid deaths will also be discussed in this section, showing the transition from predominately prescription opioid overdose deaths to heroin overdose deaths, and then the more recent transition to synthetic opioid overdose deaths. In addition, the impact of the COVID-19 pandemic on the opioid epidemic will be described, especially its role in exacerbating the opioid crisis in the United States.

With an understanding of the background of the opioid epidemic, this paper will then provide an analysis of a case study regarding the opioid crisis in the Kensington region of Philadelphia, which is one of the most impacted areas by the opioid epidemic in the country. Furthermore, this paper will then provide a detailed breakdown of the history of safe injection sites, along with their respective risks and benefits. The implications of safe injection sites for the United States will especially be emphasized in this section, along with the recent legal battle to bring safe injection sites to Philadelphia and other areas of the country. An ethical analysis of safe injection sites will then be provided from a Catholic social teaching perspective using the principles of the respect for human dignity, solidarity, the common good, and the stewardship of resources, and the relationship of safe injection sites to the harm reduction theory will also be discussed. The ethical analysis will provide an argument as to why safe injection sites should be implemented in the United States. Lastly, we will make seven recommendations for the successful and effective implementation of safe injection sites in the United States, highlighting legal, medical, educational, social, and financial aspects.

2. Opioid epidemic background

Since the 1990s, the opioid epidemic has been a devastating problem in the United States. According to the Centers for Disease Control and Prevention (CDC), between 1999 and 2019, nearly 500,000 people in the United States died from a drug overdose involving opioids, which accounts for nearly two-thirds of the total drug overdose deaths [1]. In 2019 alone, nearly 70% of the 70,630 drug overdose deaths involved opioids, and the total number of deaths from drug overdoses was four times higher in 2019 as compared to 1999 [1]. While the opioid epidemic has ravaged the United States for decades, it has not always been at the forefront of public health concerns. In 2015, Anne Case and Angus Deaton, two extremely well-known economists with the latter being a Nobel Prize winner, brought the opioid epidemic to the spotlight. While investigating morbidity and mortality rates for men and women from the CDC, Case and Deaton discovered a puzzling trend: There was a striking increase in the morbidity and mortality of middle-aged white non-Hispanic individuals between 1999 and 2013 [2]. According to Case and Deaton, between 1978 and 1998, the mortality rate for middle-aged (45–54 years old) white non-Hispanics in the United States fell by 2% per year on average [2]. However, beginning in 1999, the mortality rate for middle-aged white non-Hispanics in the United States rose by an average of half a percent a year until 2013 [2]. At the same time mortality rates were increasing for non-Hispanic whites, morbidity rates experienced a direct increase as well. In a self-reported assessment of health status from 1997 to 1999, there was a 6.7% decrease in middle-aged non-Hispanic whites from the United States reporting excellent or very good health and a related 4.3% increase in middle-aged US non-Hispanic whites reporting an increase in fair or poor health. At the same time, mortality and morbidity were increasing in the United States between 1999 and 2013 for middle-aged US non-Hispanic Whites, self-reported declines in health, mental health, increases in chronic pain and inability to work, ability to conduct daily activities of living, and clinically measured deteriorations in liver function [2]. All of these factors interestingly coincide with the increased availability of prescription opioids for pain during the 1990s [2].

The prevalence of different types of opioids has fluctuated throughout the epidemic in three distinct waves, with the first wave of the opioid epidemic involving the increased prescription of opioids. Prescription opioids, such as OxyContin and Vicodin, are frequently used to treat moderate-to-severe pain after surgery and chronic pain [3]. While the use of prescription opioids was originally intended for chronic pain from diseases, such as cancer, or for short-term use for recovering after surgery, in the 1990s medical professionals began expanding the use of opioids [4]. Physicians began to increase the long-term use of opioids in treating chronic nonmalignant medical conditions, which include conditions, such as sciatica and low-back pain [4]. Physicians were especially influenced and encouraged to increase the usage of prescription opioids, such as OxyContin, through aggressive marketing tactics from drug companies [4]. For the physicians, the message to "be proactive with pain and treat it aggressively," seemed to make perfect sense, and promoted the use of prescription opioids, such as OxyContin [4]. In 1995, OxyContin, a prescription opioid-containing the highly addictive compound oxycodone produced by Purdue Pharmaceutical, was approved by the United States Government. OxyContin was initially approved as an extended-release reformulation of oxycodone that was intended to reduce abuse and addiction [5]. Since opiates at the time were being used recreationally and it was widely believed that individuals with chronic pain needed more help, OxyContin was readily approved [6]. However, Purdue Pharmaceutical clearly lied and deceived the public by claiming that OxyContin's delayed absorption ability reduced the "abuse liability of the drug" [7].

Even though OxyContin was advertised to be less abusive than other opioids, Purdue Pharmaceutical had actually conducted a study in 1995 showing that 68% of oxycodone could be extracted from an OxyContin tablet when crushed [6]. Evidently, Purdue Pharmaceutical knew how highly addictive OxyContin was, but continued to lie about their product to increase sales. With this completely false claim of reducing potential abuse of OxyContin and the unwavering message to alleviate pain whenever possible, Purdue Pharmaceutical was able to successfully market their drug to physicians, which resulted in sales increasing from $48 million in 1996 to approximately $1.1 billion in 2000 [8]. For the next 20 years after 1995, prescription opioids, such as OxyContin, experienced a 10-fold increase in medical use. Interestingly, family medicine physicians with no expertise in pain management prescribed more opioids than any other type of physician, even pain specialists. In 2012, 18% of all opioid prescriptions were written by family medicine physicians, 15% were written by internists, and only a mere 5% were written by pain specialists [4].

Despite the rapid increase in the use of prescription opioids, such as OxyContin, in the United States beginning in the late 1990s, OxyContin was not even more effective than alternative drugs. For example, a randomized double-blind study showed that giving OxyContin every 12 hours produced comparable efficacy and safety results when treating chronic back pain as giving immediate-release oxycodone four times daily [9]. In addition, when treating patients with moderate to severe cancer-related pain, a randomized double-blind study showed that OxyContin given every 12 hours was as effective as immediate-release oxycodone given four times daily [10]. Even during the FDA's review of OxyContin in 1995, the FDA's medical review officer concluded that OxyContin had no significant advantages over immediate-release oxycodone [6]. Along with not producing a significant advantage over other alternatives, there have been no studies affirming the long-term effectiveness of OxyContin [3]. Moreover, in 2006, a Danish study with a national random sample of over 10,066 individuals that compared opioid users to non-opioid users revealed that opioid usage was significantly associated with reports of moderate/severe or very severe pain, poor self-rated health, unemployment, increased use of the health care system, and a lower quality of life [11]. With OxyContin seeming to have more negative effects on patients than positives, it is evident that its producer, Purdue Pharmaceutical, is at fault. As Oxycontin became one of the most prescribed opioids in America, it had also become one of the most abused drugs in America by 2004, resulting in an increasing number of opioid overdose deaths [12]. From 1999 to 2017, the number of opioid overdose deaths involving prescription opioids increased from 3,442 to 17,029 deaths [13].

While Purdue Pharmaceutical did nothing to inform the public about these harms, physicians are not infallible for their propagation of the opioid epidemic. Even after the addictive nature of opioids has become apparent in the past two decades, prescription opioids are still heavily utilized. In 2017 alone, over 191 million opioid prescriptions were dispensed to American patients [3]. In addition, the long-term use of prescription opioids by an individual is connected to the prescribing patterns of the original physician they encountered [14]. For example, researchers discovered that doctors they marked as "high-intensity" prescribers sent one out of four patients home with opioids, while "low-intensity" prescribers gave opioids to one out of 14 patients, and patients that saw a "high-intensity" prescriber were over 30% more likely to become long-term users of prescription opioids [14]. While physicians played a role in starting the opioid epidemic, they can continue to exacerbate it if prescription opioids continue to be prescribed at high frequencies.

Along with physicians and large pharmaceutical companies, such as Purdue Pharmaceutical, the Drug Enforcement Administration (DEA), further worsened the opioid epidemic. For example, in 2019, the Justice Department's inspector general criticized the DEA's decision to authorize manufacturers to tremendously increase the production of prescription opioid painkillers between 2003 and 2013 while opioid-related deaths in the United States surged [15]. The DEA directly oversees access to opioids, regulates opioid production quotas, and investigates illegal diversions of opioids [15]. However, the DEA failed to adequately respond to the opioid crisis as it increased the production quotas for oxycodone production by nearly 400% between 2002 and 2013, even though there was significant evidence that opioids were being abused and overprescribed [15]. The DEA also further did not capture proper data on opioid abuse and other drug trends between 2002 and 2013, thereby further handicapping their ability to properly mitigate the opioid crisis [15].

Along with the DEA, the Food and Drug Administration (FDA) further failed to deal with the opioid epidemic properly. For example, the FDA failed to use its policing powers by providing no oversight or measure of effectiveness for a safety training program that aimed to reduce the improper prescription of opioids [16]. In 2007, Congress explicitly gave the FDA the power to require the manufacturers of opioids to give training to physicians so that they could properly prescribe opioids, and the FDA was allowed to monitor the performance of these drug companies [16]. With this power, in 2011, the FDA asked producers of OxyContin and other addictive long-term opioids to pay for safety training for nearly 320,000 physicians prescribing their drugs, and also asked these entities to track the effectiveness of this training and measure other factors, such as reducing addiction, overdoses, and deaths [16].

Even though the safety program sounded good in theory, in practice it was doomed from the beginning. For example, the FDA never determined if the program worked as the opioid manufacturers were not properly collecting the right type of data, and the FDA made the critical mistake of leaving the monitoring of these safety programs in the hands of the drug manufacturers. In 2010, the FDA advisory committee of experts was aware of these potential flaws in the program's design and voted 25–10 against its implementation; however, the FDA still implemented the flawed program [16]. In addition, in 2013, a report by the inspector general of the Department of Health and Human Services showed that only 14% of the safety programs that the FDA reviewed actually met their goals, and that the FDA had no enforcement actions against companies that did not provide enough information about their safety program for it to be reviewed [16]. Hopkins researchers discovered that even though the FDA was aware of these problems after performing their own review process, they still did nothing to change it, and ultimately failed to regulate the opioid manufacturers [16]. This improper decision-making seemed to be driven by a conflict of interest between opioid manufacturers and the FDA staff responsible for opioid oversight [17]. Shockingly, the two FDA reviewers that originally approved Purdue Pharmaceutical's oxycodone application joined Purdue after leaving the FDA [17]. With a conflict of interest and the lack of action, the FDA made no progress in limiting the devasting impacts opioids were having in the United States, and ultimately played a critical role in exacerbating the opioid crisis.

While prescription opioids started the opioid epidemic in the United States, the second wave of the opioid crisis, starting in 2010, saw a rapid increase in the number of opioid overdose deaths using heroin. According to the CDC, between 2010 and 2018, opioid overdose deaths involving heroin increased by a factor of 5 from 3,036 deaths in 2010 to 14,996 deaths in 2018 [18]. In 2018, it was estimated that over 808,000 individuals had used heroin in the United States [19]. While heroin use only started to spike around 2010, it has accounted for nearly a third of all opioid overdose deaths contributing to the death of over 115,000 Americans between 1999 and 2018 [18]. Furthermore, heroin is often combined with other drugs and alcohol, leading to a greater chance of overdose. In 2013, it was reported that over 96% of heroin users use another drug, while 61% report using at least three different drugs [20]. Since heroin is typically injected using needles, heroin users are also at risk of contracting HIV, Hepatitis B, and Hepatitis C. In 2017, the CDC reported that almost 9% of the 38,738 diagnoses of HIV in the United States resulted from the injection of drugs, such as heroin [21].

Despite the devastating impact that heroin and prescription opioids have had on the United States during the opioid epidemic, the third wave of the opioid epidemic, starting in 2013, has been characterized by the development of deadly synthetic opioids. The use of synthetic opioids, such as fentanyl and Tramadol, is currently the leading cause of opioid deaths. For example, in 2018, over 31,000 people died from overdoses involving synthetic opioids excluding methadone, which was a 10% increase from 2017 [22]. From 2010 to 2013, the national rate of synthetic opioid overdose deaths was approximately 1 per 100,000 individuals, while this rate tripled from 2013 to 2015 to nearly 3.1 synthetic opioid overdose deaths per 100,000 individuals [23]. Fentanyl is particularly problematic as it is 50–100 times more potent than morphine, and is often mixed with heroin and cocaine without the knowledge of the user, as it can increase the euphoric effects [22].

With the increased usage of synthetic opioids in recent years, the opioid epidemic began to emerge as a public health crisis. On October 26, 2017, President Trump officially declared the opioid epidemic a "public health emergency" [24]. With this declaration, President Trump instructed the Health and Human Services (HHS) secretary, Eric D. Hargan to declare the opioid epidemic a public health emergency [24]. With Hargan's declaration, HHS was allowed to allocate resources and personnel to face the opioid epidemic [25]. In particular, on March 20, 2018, a National Health Emergency Dislocated Worker Demonstration Grant became available to individuals who experienced economic and workforce-related impacts caused by the opioid crisis [26]. Through providing this grant, training opportunities for skilled professions were encouraged to help those struggling with addiction to have a path back to the workforce.

While progress in mitigating the opioid epidemic has occurred in recent years, the development of COVID-19 has exacerbated the crisis. Before COVID-19, between 2017 and 2018, opioid-involved deaths decreased by 2%, prescription-opioid involved death rates decreased by 13.5%, and heroin involved deaths decreased by 4% [3]. With drug overdose mortality declining for the first time in over two decades, there was room for optimism in 2018 that the opioid epidemic was finally starting to become under control [27]. However, between 2018 and 2019, drug overdose deaths climbed once again up to 70,000 deaths [27]. This increasing trend in drug overdose deaths starting before the COVID-19 pandemic, has only continued to increase as a result of the pandemic. For example, provisional drug overdose deaths experienced an increase of 2,146 deaths from 75,696 deaths in the 12-months ending in March 2020 to 77,842 deaths in the 12-months ending in April 2020, and drug overdose deaths experienced a 3,388 death increase from the 12-months ending in April 2020 to 81,230 deaths in the 12-months ending in May 2020 [28]. What makes these numbers so alarming is that the increase of 2,146 provisional drug overdose deaths and 3,388 provisional drug overdose deaths mark the largest monthly increases since provisional 12-month estimates began to be calculated in January 2015 [28]. Even more alarming is the fact that in a 12-month period ending in September 2020, more than 87,000 Americans died from drug overdoses, which was the highest number ever recorded since the start of the opioid epidemic in the 1990s [29]. Moreover, the opioid crisis has even caused more devastation than COVID-19 in cities, such as San Francisco, where in 2020, the number of drug overdose deaths skyrocketed to 713 deaths as compared to the 257 individuals who died of COVID-19 that year [30].

With record-breaking monthly and annual surges in drug overdose deaths recorded at the start of COVID-19's declaration as a national emergency, it is clear that the pandemic has been responsible for increasing the number of drug overdose deaths. While we know this is happening, why is the pandemic increasing drug overdose deaths? First of all, while social distancing has been critical in mitigating the COVID-19 pandemic, it has also unfortunately been extremely problematic for those recovering from drugs or for individuals that use drugs. For example, access to essential treatments and community groups has been disrupted, as individuals have been instructed to stay away from others during the pandemic [27]. While recommendations to avoid individuals have been generally seen as positive, an unintended consequence of these recommendations is that they conflict with the harm reduction theory's principle of never using alone [27]. Furthermore, the pandemic is tremendously increasing the reasons people have to use drugs, such as opioids, as it has increased unemployment, feelings of loneliness and hopelessness, poverty, and a general desire to escape [27]. As people lose their jobs and experience economic and social turmoil, drugs like opioids are readily being seen as a remedy for these troublesome issues.

With the devastation that the pandemic has caused to the lives of so many individuals, it is not a surprise that heroin and synthetic opioids usage and deaths have seen tremendous rises. In a study of over 150,000 urine samples ordered by health professionals 4 months before the national emergency declaration (November 14, 2019, to March 12, 2020) and after (March 13, 2020, to July 10, 2020), fentanyl prevalence increased from 3.80% to 7.32% and heroin prevalence increased from 1.29% to 2.09% [31]. In addition, in a study of over 500,000 definite drug test results from Millennium Health in the periods before and after the national emergency for COVID-19 was declared, the national findings revealed a 31.96% increase in non-prescribed fentanyl, 19.69% for methamphetamine, 10.06% increase in cocaine, and a 12.53% increase for heroin [32]. Along with the rise in heroin and synthetic drug usage, drug overdose deaths from both of these types of opioids have increased as well. For example, between the 12-months ending in June 2019 and the 12-months ending in May 2020, the 12-month count of synthetic opioid deaths increased by over 38.4% [28]. The increase in drug overdose deaths from heroin is co-linked with drug overdose deaths, as overdose deaths from cocaine typically combined with heroin increased by 26.5% [28]. While COVID-19 has resulted in the death of over 530,000 individuals in the United States as of March 2021, the COVID-19 pandemic has only accelerated the usage and drug overdose deaths from opioids, and has pushed the opioid epidemic out of the concern of the public eye [33].

3. Case study

Philadelphia is home to the worst opioid crisis in the United States. Of the 10 most populous counties in the United States, Philadelphia has the highest overdose rate [34]. In addition, according to the Philadelphia Department of Public Health, in 2019, over 1,150 people died from drug overdoses in Philadelphia with 80% of these overdose deaths involving opioids [35]. The COVID-19 pandemic has further caused an 11% increase in the number of drug overdose deaths in Philadelphia in the first three quarters of 2020 compared to the same period in 2019 [36]. The pandemic has been especially devastating for Black and Hispanic Philadelphians, as drug overdose deaths in the first three quarters of 2020 increased by over 40.3% for Black Philadelphians and increased by 5.9% for Hispanic Philadelphians while decreasing by 7.3% for white residents [36]. These discrepancies have been accounted for by explanations of systemic racism experienced by Black and Hispanic individuals that result in less access to treatment, education, and economic resources [36].

Even though the number of deaths is troubling, the number of Philadelphians addicted to opioids is even more problematic. In 2017, the Philadelphia Department of Public Health estimated that over 75,000 of its residents are addicted to heroin and other opioids [35]. At the center of the opioid crisis in Philadelphia is one main neighborhood: Kensington. The Kensington District is home to one of the largest open-air drug markets in the United States, with buyers and sellers of heroin and other opioids roaming the streets at all hours of the day [37]. Of the 1,217 people that fatally overdosed from drugs in Philadelphia in 2017, 236 individuals fatally overdosed in Kensington alone [35]. With high levels of opioid use and fatal drug overdoses, individuals in Kensington live in a perpetual state of suffering.

One resident of Kensington, Crystal, a 34-year-old mother of three children, is a devout heroin addict. Crystal is originally from the Kensington area and had several other of her relatives that were addicted to heroin [34]. After Crystal's husband lost his job, Crystal began to utilize heroin as a coping mechanism for this loss. As Crystal and her husband divorced, she continued to abuse heroin, and eventually found herself living on the streets of Kensington without her children. Through further suffering a broken ankle, Crystal continued to use heroin as a means to escape. Crystal would constantly need to be revived by Narcan, as she was heavily addicted to heroin. The heroin in Kensington, known as "Philly dope," is especially more dangerous as it is often laced with the dangerous synthetic opioid, fentanyl [34]. The heroin is also extremely cheap in Kensington at only $5 a bag making it accessible to numerous individuals. With these factors, Crystal remained in a state of a constant dependency on heroin. When she was without it, she would experience tremendous episodes of withdrawal. Crystal described this lifestyle as "playing Russian roulette with your life," but sadly like many other Kensington residents, she was too addicted to leave this lifestyle behind [34].

Although heroin users from Kensington like Crystal can grow up in Kensington, many users are drawn to the area for its reputation as "the Walmart of heroin" [34]. One middle-aged woman named Jax migrated to Kensington to start shooting up heroin. Jax was originally a college student that started using opioids in college but eventually became a prostitute after shooting up heroin in Kensington. Jax tried to remain sober by checking herself into rehab centers but could never escape the addiction. Despite having the initial support of her boyfriend, Jax's boyfriend eventually left, leaving Jax alone to continue her heroin use. Like many women in Kensington, Jax turned to prostitution to pay for bags of heroin. Prostitutes in Kensington have been raped, tied up, and abused but are often afraid to tell police about their abuse due to their previous drug or prostitution charges. After becoming pregnant in 2009, Jax used heroin for the whole nine months, and her resulting son currently does not live with her. In 2018, Jax spent 24 days in jail, but after being released, she overdosed nine times in two weeks. Despite being saved from death on numerous occasions, Jax simply wishes that people would "just let me die" [34].

As shown through the examples of Crystal and Jax, individuals from all different backgrounds can be drawn into devastating opioid addictions. Crystal inevitably became a product of her own environment, while Jax was attracted to Kensington for its powerful opioids [34]. Although each woman experienced unique life challenges, their current lives are plagued by a constant presence of overdosing and dependence on heroin. While these women could potentially benefit from rehab centers, it is evident through cases like Jax that these rehab centers are simply not enough. Even when Philadelphia mayor Jim Kenney attempted to clean up the streets of Kensington in 2017, displaced residents from his projects continued to refuse the city's offer of treatment [34]. While rehab centers are unlikely to benefit Crystal and Jax, safe injection sites could potentially allow these individuals to slowly escape their heroin addiction, and at least mitigate the harm of abusing heroin.

4. Consequences of safe injection sites

4.1 Benefits of safe injection sites

While safe injection sites have been a recently new topic of discussion in the United States, they have benefited other countries for decades. Canada decided to become the first North American country to implement safe injection sites in 2003. Leading up to their decision, Canada had been experiencing health-related and social harms with injection drug use, especially in Vancouver [38]. In the mid-late 1990s, Vancouver's health authorities declared a public emergency after trends showed over 300 annual fatal overdoses occurring in the province of British Columbia, along with an annual new HIV infection rate of approximately 19% among local people who inject drugs [38]. As a result, the first legally sanctioned safe injection site opened in Vancouver in 2003 [38]. The Portland Hotel Society (PHS), a non-government organization, was responsible for the creation of the first sanctioned site, which initially began by members quietly building the safe injection site in a seemingly boarded up vacant building [38]. The regional health authority decided to work with PHS to open the site after its development in the form of a scientific pilot known as Insite, and the site received a federal exemption under Section 56 of the Controlled Drugs and Substances Act by the federal Health Ministry [38].

In the years since the creation of Insite, the results have been extremely promising. Since 2003, there have been over 3.6 million visits to inject drugs under the supervision of nurses, with 6,440 overdose interventions, and zero fatalities [39]. Along with preventing any overdose deaths, since 2003, there have been approximately 48,798 clinical treatment visits from users of illicit drugs. Furthermore, in the most recent data from 2019, Insite had 170,731 visits by 5,111 individuals, with 1,314 overdose interventions, and 3,158 clinical treatment interventions [39]. An average of 312 injections occurred at Insite per day in 2019, with 60% of these injections involving only opioids, 15% of injections involving stimulants only, and approximately 24% of injections involving a mix of opioids and stimulants [39]. Moreover, after the opening of Insite, the overdose mortality rate of all persons living within 500 m of the facility (70% of safe injection site users) decreased from 253 to 165 per 100,000 person-years, and one overdose death has been prevented for every 1137 users [40]. However, there was no change in overdose mortality elsewhere around the city of Vancouver, indicating the importance of Insite in inducing these changes. After the opening of Insite, there were also 67% fewer ambulance calls for treating overdoses, along with a decrease in HIV infections with an estimated 6–57 HIV infections being prevented per year in Vancouver [40].

Along with the tremendous benefits provided by Insite, the model has proved to be cost-effective. Insite's operating costs are approximately $3 million per year, but the value of the benefits of Insite is much greater [41]. Without Insite, it is estimated that yearly HIV infections in Vancouver would by 83.5 infections increase from 179.3 to 262.8 infections [41]. Using the value of 83.5 preventable HIV infections, Insite has been estimated to save approximately $17.6 million in lifetime HIV-related medical costs [41]. In addition, based on more conservative estimates, Insite has been estimated to prevent around three deaths per year and approximately 35 new cases of HIV [42]. Using these values, Insite has been estimated to have a societal benefit in excess of $6 million after yearly operating costs are factored in, thereby producing an average benefit-cost ratio of 5.21:1 [42]. Moreover, another study projected that over 54 cases of hepatitis C infections would be prevented over ten years with the presence of Insite [43]. The amount of cost averted per case was estimated to be $444,500 for each hepatitis C virus infection [43]. Through these estimates, it is clear that Insite is an effective use of public health resources when analyzing the measurable benefits of HIV infection and drug overdose deaths.

With the continued success of safe injection sites in Vancouver and around the world, it is evident that safe injection sites provide clear benefits to their communities. A summarized list of the benefits of safe injection sites is as follows:

1. Successfully managing on-site overdoses and reducing drug-related overdose death rates

2. Saving costs due to reducing disease, overdose deaths, and the need for emergency medical services

3. Reducing the risk behavior associated with HIV and other blood illnesses, such as Hepatitis C

4. Increasing the entry into substance use disorder treatment

5. Reducing the amount and frequency individuals use drugs

6. Increasing the delivery of medical and social services [44].

4.2 Bringing safe injection sites to the United States

Based on the success of Insite and other safe injection sites facilities, the number of safe injection site facilities has continued to grow. As of 2021, there are approximately 120 safe injection sites operating in ten different countries throughout the world: Australia, Canada, Denmark, France, Germany, Luxembourg, the Netherlands, Norway, Spain, and Switzerland [44]. Despite the opioid epidemic ravishing the United States, there are currently no safe injection sites in the country. However, in the past few years, several cities in the United States have been trying to change that. With the worst opioid crisis in the nation, in January 2018, the city of Philadelphia began to look toward implementing a safe injection site in the form of Comprehensive User Engagement Sites (CUES) [45]. CUES was developed with Insite as a model and adopted many of its services, such as monitoring user injection of drugs for on-site overdose care, recovery/detoxification services, referrals to treatment, and other health services, such as wound care, immunizations, and pregnancy tests [46]. CUES was also structurally similar to Insite by providing a reception area for PWID and giving each PWID a card with an anonymous identification number [46]. The purpose of the card is not to collect personal information, but rather to provide PWID the address and phone number of the facility, phone number of counseling and rehabilitation programs, and also contacts to emergency services in case of an overdose [46]. At the reception center, PWID are also asked where they heard about the site from, their age and ethnicity, along with their interest in services, such as rehabilitation, psychiatric services, wound care, and clean needle exchanges [46]. After the reception, PWID can safely inject at separate benches in a large room under the supervision of healthcare professionals, and can then briefly relax in a separate lounge area for 30 minutes. During the lounge period, educational material, counseling, and rehabilitation services are offered to PWID [46]. Unlike Insite, CUES offers additional services, such as fentanyl screenings, Hepatitis-C/HIV screenings, a needle exchange program, Narcan distribution and education, early education, and counseling for rehabilitation and detoxification done by individuals in recovery who have actually experienced drug addiction [46].

The services provided by CUES also have extremely promising impacts on population health. For example, in December 2017, SCF-averted HIV infections were estimated to range from 1 to 18 annually with the low range assuming a receptive needle sharing of 2% with the upper range assuming a receptive needle sharing of 28.3% [47]. In addition, SCF-averted hepatitis C virus (HCV) infections were estimated to range from 15 to 213 annually assuming the same receptive needle sharing ranges as used in the HIV estimates [47]. While the low ranges are possible for both averted HIV and HCV infections, the Philadelphia Department of Health estimates that the actual rate of needle sharing is closer to 28.3%, thereby suggesting that true values for averted infections should be closer to the upper limits [47]. Along with HIV and HCV infections, the estimated number of annual overdose deaths averted within 500 meters of a SCF would be between 27 and 48 deaths annually, while the number of averted overdose deaths from opening a SCF would be between 24 and 76 deaths annually [47].

Along with the health impacts, the estimated financial impacts of SCFs are quite striking. For example, the estimated total value of overdose deaths averted in Philadelphia would be between $12,462,213 and $74,773,276 annually [47]. In addition, the estimated annual savings due to SCF skin and soft tissue infection (SSTI) reductions would be between $1,512,356 and $1,868,205 annually [47]. Furthermore, estimates predict that SCF will result in savings due to a reduction in emergency room visits, hospital stays, and ambulance calls. For example, the estimated annual savings due to SCF reducing ambulance calls for an overdose is $123,776 [47]. The estimated annual savings from keeping PWID out of the emergency room is $280,683, while the estimated annual savings on hospitals for PWID who overdose would be $247,941 [47]. With these estimates, SCFs are expected to produce tremendous benefits to the health and financials of individuals in the city of Philadelphia.

With the promising impacts of safe injection sites in Philadelphia, numerous individuals in Philadelphia saw developments, such as CUES as an opportunity to successfully combat the opioid epidemic. In 2018, the Mayor of Philadelphia, Jim Kenney, and Philadelphia District Attorney Larry Krasner, both strongly supported the implementation of CUES. Instead of combating drug addiction through the criminalization of drugs by claiming a "War on Drugs," both Kenney and Krasner wanted to approach drug addiction from more of a harm reduction standpoint [48]. Through the opioid epidemic driving violence in Philadelphia and stressing the EMS system, both Kenney and Krasner realized that immediate action was needed that focused on saving lives "in a way that research has shown to be successful." [48]. For Kenney and Krasner, CUES provides not only a place where PWID can be safely injected under supervision, but also provides a direct link to treatment, resources for housing and meals, and the ability to save lives [48]. Krasner also expressed that he would not prosecute anyone involved in bringing safe injection sites to Philadelphia [48]. Moreover, safe injection sites, such as CUES, have the support of Philadelphia's Health sector. In 2018, Dr. Thomas Farley, Philadelphia's Health Commissioner and co-chair of the city's opioid task force, also expressed his support for safe injection sites [49]. Farley described how there are "many people who are hesitant to go into treatment, despite their addiction, and we do not want them to die" [49]. Instead of letting individuals die, Farley continued to reiterate the sentiment that safe injection sites save lives and connect individuals to treatment [49]. Based on the support of numerous public officials, as of early 2018, the message for safe injection sites was clear: CUES have the greenlight to be implemented. Although, instead of paying or operating any of the safe injection sites, the city of Philadelphia would simply be a facilitator and connector with the organizations that provided addiction services [49].

4.3 Legal barriers to safe injection sites in the United States

With the city of Philadelphia not directly funding safe injection sites, other organizations became responsible for the implementation of safe injection sites. Within a year, a privately funded, 501 tax-exempt, Philadelphia nonprofit corporation named Safehouse, attempted to address the opioid epidemic crisis by its plan to open the United States' first safe injection site in Philadelphia [50]. However, Safehouse's plan was met with immediate legal resistance. In February 2019, Bill McSwain, the U.S. Attorney for the Eastern District of Pennsylvania filed a civil lawsuit that asked a federal court to declare supervised consumption sites, otherwise known as safe injections sites, illegal under 21 U.S. Code 856, which is a section of the Controlled Substances Act known as the "Crack House" Statute [50]. The Federal Comprehensive Drug Abuse Prevention and Control Act of 1970, known as the Controlled Substances Act, had the main goal of improving the manufacturing, importation, exportation, distribution, and dispensing of controlled substances [51]. To create a "closed system" for controlled substances, the Controlled Substances Act further mandated that manufacturers, distributors, and dispensers of controlled substances must register with the Drug Enforcement Administration (DEA) [51]. In addition, the Controlled Substances Act categorized controlled substances into five schedules based on their abuse potential to further aim to mitigate the potential harms of these substances [51].

In terms of safe injection sites, the Controlled Substances Act, has tremendous legal implications. For example, Section 856(a) of the Controlled Substances Act, otherwise known as the "Crack House Statute," contains a potential barrier to establishing safe injection sites. Subsection (a) regarding the unlawful acts of Section 856 of the Controlled Substances Act is as follows:

Except as authorized by this subchapter, it shall be unlawful:

1. Knowingly open, lease, rent, use, or maintain any place, whether permanently or temporarily, for the purpose of manufacturing, distributing, or using any controlled substance;

2. Manage or control any place, whether permanently or temporarily, either as an owner, lessee, agent, employee, occupant, or mortgagee, and knowingly and intentionally rent, lease, profit from, or make available for use, with or without compensation, the place for the purpose of unlawfully manufacturing, storing, distributing, or using a controlled substance [52].

Based on the Crack House Statute, it is illegal to run a facility where controlled substances, such as opioids, are used. With this statute, operating these facilities is considered a federal crime and individuals can be subject to a felony with imprisonment of up to 20 years, and hundreds of thousands of dollars in fines [52]. With McSwain's civil lawsuit against Safehouse, it now became up to the federal courts to decide if safe injection sites indeed violate the Crack House Statute. McSwain in conjunction with the Justice Department argued that the plain language of the law was clear, and if advocates of Safehouse wanted to change the law to open safe injection sites, they would have to lobby Congress [53]. On the other hand, advocates for Safehouse argued that the Crack House Statute does not apply to safe injection sites as the law only intended to impact the owners and tenants of drug dens at a time when the crack-cocaine epidemic was at its height [53]. Despite McSwain's appeal, in October 2019, U.S. District Judge Gerald A. McHugh did not believe the Crack House Statute applied to safe injection sites, such as Safehouse, as he did not believe that lawmakers had safe injection sites in mind when creating the Controlled Substances Act [53]. McHugh further argued that the goal of safe injection sites, such as Safehouse, is to "reduce drug use, not facilitate it, and that it was up to Congress to amend the statute if it wanted to deem safe injection [53].

After this initial victory defending the implementation of safe injection sites, the battle over the legality of safe injection sites has continued. On February 25, 2020, McHugh issued a Final Declaratory Judgement for Safehouse which declared that the proposed safe injection site did not violate any of the federal drug laws [50]. Safehouse asked for this judgment to be assured that they could proceed with their plans to open a proposed safe injection site [54]. With this reaffirmation of the legality of safe injection sites, Safehouse was ready to bring the first safe injection site inside the Constitution Health Plaza at the corner of Broad and McKean Street in South Philadelphia [54]. However, McSwain adamantly continued his argument against the implementation of safe injections sites under the current law, and appealed McHugh's decision to the U.S. Court of Appeals for the Third Circuit a few days after McHugh's final judgment [55]. McSwain also asked for an Emergency Motion of Stay, which would prevent safe injection sites from opening until the Third Circuit court made its decision [50]. Along with McSwain, community residents in South Philadelphia resisted the idea of safe injection sites. Residents argued that crime would be brought to their neighborhoods, and that safe injection sites were not going to solve the opioid epidemic [56].

Despite the resistance, Safehouse did not want to slow the progress they had made. On March 10, 2020, in response to McSwain's order for a stay to be issued, Safehouse requested that a stay not be issued on the case [50]. However, during the following week, the city of Philadelphia went into lockdown due to the COVID-19 pandemic, resulting in the closure of courts across the Philadelphia area [57]. Along with the COVID-19 pandemic, protests against police brutality swept across the nation after the killing of Mr. George Floyd in late May 2020 [58]. With city and country officials being overwhelmed with the challenges associated with COVID-19 and the outrage over police brutality, McHugh ultimately decided to approve McSwain's emergency stay on June 25, 2020 [58]. Even though McHugh did not change his original verdict on safe injection sites, he believed it was the "wrong moment for another change in the status quo" [58]. After a long delay, a three-member panel of the Third Circuit Court of Appeals eventually started hearing oral arguments on November 16, 2020 [50]. At the conclusion of the case hearing, the Third Circuit Court of Appeals agreed with the Government that supervised safe injection sites are illegal under federal law on January 13, 2021 [59]. In a 2-1 decision, the appeals court decided to change to the decision of the federal district court as it argued that Safehouse "knows and intends that visitors to its consumption room will have a significant purpose of using illegal drugs" [59]. With this decision, Safehouse suffered a massive blow, as opening safe injection sites became declared to be illegal in the United States. On February 24, 2021, Safehouse filed a Petition for Rehearsing En Banc, which requested a rehearsal in front of the entire panel, but this request was denied on March 24, 2021, despite three judges issuing strong dissents to the denial [50]. Safehouse still has options to appeal to the Supreme Court of the United States.

With the promise of a ruling on safe injection sites going to the Supreme Court of the United States, several states have continued to lobby for state laws that support the implementation of safe injection sites. For example, after first introducing the idea of Harm Reduction Centers in 2019, lawmakers revised two versions of a bill known as H 5245 and S 0016, which focus on the creation of a harm reduction center advisory committee and pilot program [60]. The Senate version of the bill, S 0016, which was passed on February 23, 2021, would result in the establishment of an advisory committee that makes recommendations to the state's Health Department Director on the regulation of safe injection sites. Unlike the house version, the Senate version of the bill provides liability protection that prevents "property owners, managers, employees, volunteers, clients, or participants, and state, city, or town government employees acting in the course and scope of employment" from being arrested or prosecuted [60]. If the house version of the bill, H 5245, is passed by the house and approved by Rhode Island Governor Gina Raimondo, Rhode Island could be the first state to legalize safe injection sites [60].

Along with Rhode Island, on March 2, 2021, the New Mexico House passed HB 123, a bill authorizing counties to establish and operate OPPs (Overdose Prevention Programs) following guidelines that will be created by the New Mexico Department of Health by October 1 [61]. In the argument for the creation of OPPs, lawmakers took a similar legal stance to the proponents of Safehouse and argued that the purpose of these sites is to protect the health of people who do drugs, not to facilitate drug use [61]. If the legislation passes through the New Mexico State Senate, OPPs will be established in New Mexico. Despite the ruling against Safehouse in January 2021, state legislatures remain optimistic that their versions of safe injection sites will be federally legal, as President Biden has strongly emphasized the key component of harm reduction in his National Drug Control Policy. The Biden-era Office of National Drug Control Policy (ONDCP), places a great emphasis on "confronting racial equity issues related to drug policy" and "enhancing evidence-based harm reduction efforts" [62]. With strong support for harm reduction in combating drug abuse by the President, the legal approval of safe injection sites seems to be on the horizon.

4.4 Potential risks of safe injection sites

While safe injection sites carry tremendous potential benefits, there are also some important risks. Outside of the legal issues previously discussed, one main potential risk of safe injection sites proposed by critics is that they can encourage drug use and disincentive drug cessation by sending out the message that drug use is legally tolerable [46]. One study on Insite in Vancouver measured changes in drug use among PWID in Vancouver both one year before and after Insite was opened. This study showed that there were no significant differences in rates of relapse into injected drug use (17% versus 20%), or stopping injected drug use (17% versus 15%) [63]. In addition, the study showed that there were no significant differences in crack cocaine smoking (12% versus 14%), rates of stopping methadone use (11% versus 7%) or starting methadone use (13% versus 11%), and rates of stopping binge drug use (58% versus 63%) [63]. The opening of Insite clearly did not significantly affect community drug use or drug cessation. In addition, another study examined drug injection cessation between December 2003 and June 2006 in Vancouver after the opening of Insite with a sample of over 1090 participants, and it was found that 46% of participants entered into treatment [64]. Based on this study, drug injection cessation does not appear to decrease or be disincentivized as a result of opening up safe injection site facilities, such as Insite, as PWID entered treatment to stop drug use.

Along with potential increases in drug use and the disincentivizing of drug cessation, another potential risk of safe injection sites voiced by critics is that these sites increase crime and neighborhood disorder. In a study comparing annual periods before and after the opening of Insite in Vancouver, it was found that rates of drug trafficking, assaults, and robbery were similar after Insite's opening [65]. Based on this study, neighborhood crime rates have not significantly increased or decreased as a result of opening safe injection sites, such as Insite. Similar to the trends in crime rates, there has been no evidence to show that safe injection sites increase neighborhood disorder. In a study comparing public injection drug use and public syringe disposal before and 12-weeks after the opening of Insite, there were statistically significant reductions in the daily mean number of injecting drug users injecting in public from approximately 4.3 users to 2.4 users and significant reductions in publicly discarded syringes from a daily average of 11.5 to 5.3 syringes [66]. This study also showed that there was a statistically significant reduction in injection-related litter for Insite as the pre/post daily mean count of injection-related litter decreased from 601.7 to 305.3 [66]. Based on this study, safe injection sites do not appear to increase neighborhood disorder. Moreover, critics of safe injection sites have also feared that these sites will promote drug tourism from outside of the community [46]. Despite the evidence showing that increases in crime rates and neighborhood disorder are unlikely, no studies are currently available that have analyzed the potential problem of drug tourism. However, scholars have pointed out that the majority of PWID users are residents in the area surrounding the safe injection Insite, so it would be unlikely that after the implementation of Insite people would travel long distances just to use drugs in a dangerous and impoverished area of Vancouver [46].

While crime rates, neighborhood disorder, and drug tourism seem unlikely to increase as a result of safe injection sites, continued potential risk of safe injection sites is the lack of community support for them. For example, South Philadelphia residents felt "blindsided" by Safehouse's decision to plan a safe injection site in South Philadelphia, and accused Safehouse of never soliciting the community's support [49]. Many residents located near potential safe injection sites fear the potential for increases in crime and neighborhood disorder with some completely rejecting the notion that safe injection sites can even mitigate the opioid epidemic. However, one study analyzing community perceptions of neighborhood disorder 5 years after the opening of the Uniting Medically Supervised Injecting Centre (MISC) in Australia, showed that the fears of residents seem to dissipate over time [67]. In this study, business owners and residents in the surrounding area of MISC noted that they had witnessed lower instances of public injecting and publicly discarded injecting equipment [67]. Over 90% of participants in this study also reported at least one advantage of MISC in their area [67]. Even though community support of safe injection sites tends to increase over time, it appears to be important to consult the community of proposed safe injection site locations to educate the members about the benefits and potential risks. This education could potentially make residents more understanding of safe injection sites from the start of their implementation.

According to critics, another potential risk of safe injection sites is that they will fail as PWID do not even want to quit injecting drugs to begin with. Scholars have argued that critics who have this viewpoint presuppose that PWID are "completely void of any desire to quit the drug habit, or that the PWID evaluate their desire to stay addicted so positively that the desire to quit is an insignificant element in their decision-making process" [46]. In a scenario like Insite where drug addiction is strong for users, it is false that PWID are completely void of any desires as addicted people experience a conflict between the desire to stay addicted and the desire to do away with drugs and start a normal life [46]. In addition, scholars believe that the view that the desire to quit is an insignificant element for PWID is false, as it implies that PWID enjoy a low quality of life along with the drug-induced euphoria they experience [46]. It is unreasonable to think that PWID enjoy their low quality of life filled with homelessness and violence in many cases. In a study of 42 PWID evaluating their perceptions of safe injection sites, several participants felt that safe injection sites would improve their neighborhoods through a lessened community exposure to drug use and less injection equipment on the streets [68]. Clearly, many PWID want to escape their life of drugs but need a facility, such as a safe injection site, to help overcome their addiction.

Along with the risk of safe injection sites failing based on the belief that PWID do not want to stop injecting drugs, safe injection sites also have financial implications. Unlike Canada where the operating costs of Insite have been provided by the British Columbia Ministry of Health Services and additional funding has been provided by Health Canada, cities in the United States are not currently funding safe injection sites [69]. Since the city of Philadelphia is not funding safe injection sites, many individuals rightfully question how these facilities will be funded and maintained. Before the potential opening of Safehouse in February 2020, Safehouse only had $200,000 in the bank, which was tremendously less than the estimated annual $1 million operating costs of the facility [54]. However, recently in March 2021, the $1.9 trillion American Rescue Plan Act was enacted, which allocated $4 billion to address the overdose crisis in America and face the challenges of substance use disorder and mental health [70]. Of the $4 billion, $30 million was allocated to "support community-based overdose prevention programs, syringe service programs, and other harm reduction services" [50]. This commitment to harm reduction may set the stage for funding for safe injection sites in the long term, although the funding of safe injection is still currently in the hands of private donors in the case of safe injection sites, such as Safehouse.

5. Ethical analysis from a catholic social teaching perspective

The development of safe injection sites as a solution to the devasting opioid crisis in the United States has caused a tremendous debate amongst individuals from a wide variety of backgrounds. As previously discussed, the advantages of safe injection sites are extensive. Safe injection sites have been shown to successfully manage on-site overdoses, reduce drug-related overdose death rates, reduce the risk behavior associated with HIV and Hepatitis C, increase the delivery of medical and social services, and increase the number of individuals entering into substance use disorder treatment. In addition, further advantages of safe injection sites include reducing the amount and frequency individuals use drugs, along with saving costs via reducing the prevalence of diseases, overdose deaths, and the need for emergency medical services. While safe injection sites, such as Insite, have clearly been shown to be cost-effective and successful in combating the opioid crisis in Vancouver, safe injection sites are not without their concerns. Critics of safe injection sites highlight concerns of increased drug use, crime, neighborhood disorder and drug tourism, reduced drug cessation, and the notion that safe injection sites will fail as PWID do not even want to quit drugs to begin with. Furthermore, critics have concerns about the potential lack of community support, how safe injection sites will be funded, and the legality behind them.

While these concerns have been previously discussed, one notion of the argument that seems to be missing is the ethical implications of safe injection sites. Are safe injection sites even ethical to begin with? Clearly, it would not be advisable to support a project that is unethical in nature, even despite the end result of the means. Even if safe injections sites provide the end result of mitigating the opioid crisis, they simply cannot be unethical in nature if they are to be successfully instituted in the United States. To determine if safe injection sites are ethical, the Catholic social teaching principles of respect for human dignity, solidarity, the common good, and the stewardship of resources will be applied to safe injection sites. The relationship of safe injection sites to the harm reduction theory will also be elaborated on to evaluate the ethics of safe injection sites.

5.1 The principle of human dignity

The principle of human dignity is known as the foundational principle for Catholic social teaching. The United States Conference of Catholic Bishops describes the principle of human dignity as the following:

> *"Every human being is created in the image of God and redeemed by Jesus Christ, and therefore is invaluable and worthy of respect as a member of the human family"* [71].

As expressed in this principle, every human being, from their conception to their death has inherent dignity and right to life based on that dignity. In practice, every human being, regardless of their personal race, religion, sex, age, national origin, sexual orientation, economic status, health, intelligence, or any other characteristic that differentiates individuals from one another, is worthy of respect [72]. Despite differing characteristics between individuals, simply being a human being confers one this dignity. As a result of this dignity, the human person is always seen as an end in the Catholic view, not as a means. The presence of human dignity further guarantees every human person a claim of membership in a community known as the human family.

Applying the respect for human dignity to safe injection sites, all individuals, especially PWID, have an inherent value to their lives that must be respected. Despite using drugs, PWID are equal to every other human person in that their dignity must be safeguarded at all times. PWID are part of the vulnerable group of individuals across the world known as the invisible and dispensable minority. As a result, PWID are often stereotyped as less valuable members of society, and their dignity is overlooked. Safe injection sites respect the dignity of PWID by directly reducing drug use, drug overdose deaths, and diseases, all of which if left unattended would directly threaten the inherent value of each PWID's life. By helping to encourage PWID into substance use disorder treatment and giving them medical and social services, safe injections sites actually enhance the respect for human dignity for PWID by protecting these individuals' lives. Safe injection sites ultimately accept PWID for who they are and where they are at the present moment.

5.2 The principle of solidarity

The principle of solidarity at its core functions is to promote peace and justice for all, along with protecting the common good. The United States Conference of Catholic Bishops describes the principle of solidarity as the following:

> *"Catholic social teaching proclaims that we are our brothers' and sisters keepers, wherever they live. We are one human family … . Learning to practice the virtue of solidarity means learning that 'loving our neighbor' has global dimensions in an interdependent world"* [71].

As evident by this principle, every human being is part of one family known as the human family. Despite our national origins or any social, political, or economic barriers that separate us, every human being is interconnected with one another. Promoting solidarity for all involves a commitment to every human being that you will seek peace and justice for everyone and promote the common good. The principle of solidarity is grounded in the Gospels' calls for human beings to be peacemakers, and Pope Paul VI has elaborated on this notion by stating that "if you want peace, work for justice" [73]. Solidarity ultimately facilitates peace, justice, and the common good for all.

Despite this sense of togetherness across the globe, solidarity has been threatened by the lack of concern for the poor and the most vulnerable individuals in society. The United States Bishops have even promoted the notion that more attention should be provided to "the needs of the poor, the weak, and the vulnerable, in a debate often dominated by more powerful interests" [71]. In terms of PWIDs, these individuals are part of the vulnerable, poor, and weak in society. PWIDs often lack social support and financial resources to overcome their addiction to the drug, leading them to continue in a constant cycle of addiction. Safe injection sites promote solidarity by emphasizing a preferential option for the poor and vulnerable by giving them adequate care and resources to overcome their addiction. While others may look down upon PWIDs, safe injection sites graciously accept them for the human beings they are and seek to invoke unity to overcome their addiction together. Safe injection sites are a direct commitment to the global community to reduce human suffering and ensure the value of human dignity for all.

5.3 The principle of the common good

The principle of the common good is an essential ethical imperative in Catholic social teaching. The United States Conference of Catholic Bishops describes the principle of the common good as the following:

> *"The common good is understood as the social conditions that allow people to reach their full human potential and to realize their human dignity"* [71].

Based on this definition, the common good is inevitability linked to solidarity and human dignity. The social conditions that promote the common good presume the "respect for the person," "the social well-being and development of the group," and the maintenance by a public authority that promotes the ideals of peace and justice [72]. Without the foundational pillars of solidarity and human dignity, there would be no common good. In addition, the common good implies the good of all and every individual, which should be sought together collectively. However, this notion of the common good directly contrasts with utilitarianism's promotion of the greatest good for the greatest number. This utilitarian belief accepts the presence of a minority group always being injured, while the common good promotes equal treatment for all human persons. Since we are a global family of brothers and sisters, we all have a right and a duty to participate in society to seek the common good and well-being of all, especially for the poor and the most vulnerable individuals in society [71]. The Catholic Church emphasizes that the major role of governments and institutions is to protect human life and human dignity and to promote the common good [71].

With the presence of global interdependence in contemporary times, the principle of the common good emphasizes the "need for international structures that can promote the just development of the human family across regional and national lines" [72]. Since it is the government's duty to promote the common good for all, including PWID, the creation of safe injection sites would support the goal of addressing the needs of the most vulnerable and poor in society. As a community with inherently valuable lives, we have a duty to support PWID. Reducing drug use, drug overdose deaths, and disease are all clear components of the common good that promote the inherent value of human life and are a preferential option for the poor and vulnerable. The common good could easily be promoted by governments in this case with the development of facilities, such as safe injection sites, that seek to combat the devastation of drug use, which affects the human person, and as a result, the global human community.

5.4 The principle of stewardship

The principle of stewardship is one that emphasizes the value of management, not ownership. The United States Conference of Catholic Bishops describes the principle of stewardship as the following:

> *"The Catholic tradition insists that we show our respect for the Creator by our stewardship of creation"* [71].

Using this definition of stewardship, it is evident that a key role in Catholic social teaching is to protect the people and planet that God created. Since we are all created in the image of God, we not only have a moral responsibility to the environment, but also to our personal talents, personal health, and our use of the private property [72]. The principle of stewardship, therefore, contains the stewardship of resources, which in itself emphasizes a just allocation of resources in the world. This requirement for the just allocation of resources has tremendous implications for safe injection sites. For example, safe injection sites fulfill the duty of the stewardship of resources by providing a just allocation of healthcare resources. Safe injections sites, such as CUES discussed above, provide health care services, such as wound care, pregnancy tests, immunizations, recovery/detoxification services, referrals to treatment, and on-site overdose care to PWID, who are some of the most vulnerable individuals in society. By justly allocating healthcare resources to PWID, the principle of the stewardship of resources is effectively and adequately promoted. Furthermore, safe injection sites, such as CUES, provide the vulnerable with equitable access to food and housing, which PWID would not have adequate access to without these facilities. Providing basic necessities and rights to individuals, such as the right to healthcare and safe injection sites, are improving the lives of individuals, which in turn improves the overall well-being of the interconnected global community. Moreover, the more general principle of stewardship is applicable to safe injection sites through their ability to reduce waste in the environment. As previously described, safe injection sites reduce the number of drug injecting equipment waste present in the community, which further fulfills the goal of stewardship in promoting respect for God's creation.

5.5 Ethical justification using the harm reduction theory

The driving ethical force behind the push for safe injection sites like CUES to be made available as a viable option for PWID is their potential to be used under the harm reduction idea. Harm reduction is an approach focused on minimizing the negative results that go hand-in-hand with drug abuse [74, 75]. Harm reduction techniques have both a medical and ethical impact on the individual and society as a whole. Harm reduction techniques accept the individuals as they are, while also tailoring that person's treatment to fit his or her needs [76]. Furthermore, there are certain principles that are quintessential to an understanding of harm reduction, as listed by the Harm Reduction Coalition:

- Accepts, for better and/or worse, that licit and illicit drug use is part of our world and chooses to work to minimize its harmful effects rather than simply ignore or condemn them.

- Understands drug use as a complex, multi-faceted phenomenon that encompasses a continuum of behaviors from severe abuse to total abstinence and acknowledges that some ways of using drugs are clearly safer than others.

- Establishes quality of individual and community life and well-being–not necessarily cessation of all drug use–as the criteria for successful interventions and policies.

- Calls for the non-judgmental, non-coercive provision of services and resources to people who use drugs and the communities in which they live to assist them in reducing attendant harm.

- Ensures that drug users and those with a history of drug use routinely have a real voice in the creation of programs and policies designed to serve them.

- Affirms drugs users themselves as the primary agents of reducing the harms of their drug use and seeks to empower users to share information and support each other in strategies that meet their actual conditions of use.

- Recognizes that the realities of poverty, class, racism, social isolation, past trauma, sex-based discrimination, and other social inequalities affect both people's vulnerability to and capacity for effectively dealing with drug-related harm.

- Does not attempt to minimize or ignore the real and tragic harm and danger associated with licit and illicit drug use [74].

The CUES' ability to allow PWID to have a safe environment to inject drugs gives itself the potential to be used as a harm reduction agent in and of itself. Furthermore, many individuals who die from opiate overdoses, such as heroin, did not receive the necessary medical treatment in time to save them; allowing PWIDs access to the CUES could possibly save many preventable deaths. If we, as a society, value human life as sacred, we must find a way to prevent these deaths. The CUES program, such as Insite in Vancouver, supervised by trained medical personnel as a harm reduction agent could present a viable alternative to address the growing heroin addiction epidemic and save thousands of lives. The heroin epidemic is growing, fatal overdoses are increasing, and people are becoming more and more frustrated by legal and political barriers to new forms of treatment being put in place to stop this problem. As shown above, SIFs like Insite have been shown to decrease heroin abuse, disease, and mortality rate in Canada and Europe. In the United States, overdoses have led to 45,000 opioid overdose deaths in a 12-month period that ended in September 2017. This number is unacceptable by any standards [77]. Therefore, the harm reduction initiatives like Insite and a CUES must be introduced.

6. Conclusion

Safe injection sites are emerging as both an effective and ethical solution to the opioid crisis in the United States. However, to deal with the various concerns about safe injection sites along with the potential for additional benefits, this article seeks to make a variety of recommendations concerning the implementation of safe injection sites. These seven recommendations include legal, medical, educational, social, and financial implications that seek to make safe injection sites as effective as possible in the United States.

1. Legal: Mobile and portable units like a van could be used as safe injection sites to overcome the Crack House Statute. Even McSwain himself, a staunch critic of safe injection sites, acknowledged that mobile units would not be a violation of the Crack House Statute [78]. Since mobile and portable units do not pertain to real estate, McSwain does not think the Crack House Statute applies to them. Prevention Point Philadelphia already uses vans effectively for their clean needle exchange program.

2. Educational: Provide medical residents, physician assistants, nurse practitioners, and medical students the opportunity to do rotations at safe injection sites. This would allow these medical professionals to be educated on communication skills with vulnerable populations, which is often not addressed in medical education programs. The opportunity to do rotations would also give medical professionals a better sense of cultural sensitivity, further allowing these individuals to have a more holistic understanding of medicine.

3. Medical: Mandatory fentanyl testing at all safe injection site facilities to protect PWID. As discussed previously, fentanyl is over 50–100 times more potent than morphine and is becoming increasingly prevalent in cities such as Philadelphia. If we really care about PWID and want to respect their human dignity and their lives, then we should seek to mitigate the extremely dangerous risk of injecting opioids laced with fentanyl.

4. Social: Safe injection sites should provide social support services such as housing, employment, and job development counseling. This would help PWID

escape their addiction lifestyle, and give them skills and knowledge to be successful and productive members of society.

5. Social: Create an advisory board to combat the concern of the lack of community involvement in the implementation of safe injection sites in various communities. This advisory board would be a diverse group consisting of medical professionals, community leaders, social workers, addiction counselors, and clergy.

6. Medical/Social: Preferably utilize addiction counselors who have been former addicts as they have a deeper relationality to PWID, which will allow them to treat PWID for who they are and where they are at in the present moment. These addiction counselors have walked the walked and talked the talk, which will enhance the rehab counseling experience.

7. Financial: No city funding should be put toward safe injection sites, as the use of tax dollars for safe injection sites would likely be extremely controversial. Instead, we recommend that safe injection sites rely on funding from grants and nonprofits, and potential partnerships with one to two local health systems. In Philadelphia, the five main health systems include Jefferson Health, The University of Pennsylvania Health System, Tower Health, Temple Health, and Trinity Health. Safe injection sites have been proved to be cost-effective and save medical resources by reducing the number of emergency room visits and the need for emergency services. As a result, partnering with safe injection sites could save healthcare systems money, and be a cheaper alternative than the expensive use of emergency medical resources.

With these recommendations, safe injection sites could legally and effectively open in cities across the United States with the aim of mitigating the devastation that the opioid epidemic has caused in the country for the last three decades.

Author details

Peter A. Clark* and David Grana
Institute of Clinical Bioethics, Saint Joseph's University in Philadelphia, USA

*Address all correspondence to: pclark@sju.edu

References

[1] Opioid Overdose: Opioid Basics. Centers for Disease Control and Prevention, Centers for Disease Control and Prevention. 2021. Available from: www.cdc.gov/drugoverdose/epidemic/index.html

[2] Case A, Deaton A. Rising morbidity and mortality in midlife among White Non-Hispanic Americans in the 21st century. PNAS, National Academy of Sciences. 2015. Available from: www.pnas.org/content/112/49/15078

[3] Understanding the Epidemic. Centers for Disease Control and Prevention, Centers for Disease Control and Prevention. 2021. Available from: www.cdc.gov/drugoverdose/epidemic/index.html

[4] Friedman RA. How doctors helped drive the addiction crisis. The New York Times. 2015. Available from: www.nytimes.com/2015/11/08/opinion/sunday/how-doctors-helped-drive-the-addiction-crisis.html

[5] Bulloch M. How oxycodone has contributed to the opioid epidemic. Pharmacy Times. 2018. Available from: www.pharmacytimes.com/contributor/marilyn-bulloch-pharmd-bcps/2018/08/how-oxycodone-has-contributed-to-the-opioid-epidemic

[6] OxyContin History. American Addiction Centers, American Addiction Centers. 2018. Available from: luxury.rehabs.com/oxycontin-rehab/history/#:~:text=OxyContin%20is%20a%20strong%20analgesic

[7] United States Court of Appeals, Second Circuit: Purdue Pharma v. Kentucky. FindLaw. 2013. Available from: caselaw.findlaw.com/us-2nd-circuit/1619840.html

[8] Van Zee, Art. The promotion and marketing of oxycontin: Commercial Triumph, Public Health Tragedy. National Center for Biotechnology Information (NCBI), American Journal of Public Health. 2009. Available from: www.ncbi.nlm.nih.gov/pmc/articles/PMC2622774/

[9] Hale ME et al. Efficacy and safety of controlled-release versus immediate-release oxycodone: Randomized, double-blind evaluation in patients with chronic back pain. PubMed, The Clinical Journal of Pain. 1999. Available from: pubmed.ncbi.nlm.nih.gov/10524470

[10] Kaplan R et al. Comparison of controlled-release and immediate-release oxycodone tablets in patients with cancer pain. PubMed, Journal of Clinical Oncology. 1998

[11] Eriksen J et al. Critical issues on opioids in chronic non-cancer pain: An Epidemiological Study. PubMed, The Journal of the International Association for the Study of Pain. 2006

[12] Cicero TJ et al. Trends in abuse of oxycontin and other opioid analgesics in the United States: 2002-2004. PubMed, The Journal of Pain. 2005. Available from: pubmed.ncbi.nlm.nih.gov/16202959/

[13] Overdose Death Rates. National Institute on Drug Abuse, National Institute on Drug Abuse. 2021. Available from: www.drugabuse.gov/drug-topics/trends-statistics/overdose-death-rates

[14] Hoffman J. Long-term opioid use could depend on the doctor who first prescribed it. The New York Times. 2017. Available from: www.nytimes.com/2017/02/15/health/long-term-opioid-use-doctors-prescriptions.html

[15] Fortin J. Let opioid production surge as crisis grew, Justice Dept. Says. The New York Times. 2019. Available from:

www.nytimes.com/2019/10/01/us/dea-opioid-crisis.html?auth=login-email&login=email

[16] Goodnough A, Sanger-katz M. As tens of thousands died, F.D.A. failed to police opioids. The New York Times. 2019. Available from: www.nytimes.com/2019/12/30/health/FDA-opioids.html

[17] Kolodny A. How FDA failures contributed to the opioid crisis. American Medical Association, Journal of Ethics: American Medical Association. 2020. Available from: journalofethics.ama-assn.org/article/how-fda-failures-contributed-opioid-crisis/2020-08

[18] Heroin. Centers for Disease Control and Prevention, Centers for Disease Control and Prevention. 2021. Available from: www.cdc.gov/drugoverdose/opioids/heroin.html

[19] Find Help: ATOD. SAMHSA, U.S. Department of Health and Human Services. 2021. Available from: www.samhsa.gov/find-help/atod

[20] Jones CM et al. Vital signs: Demographic and substance use trends among heroin users—United States, 2002-2013. Centers for Disease Control and Prevention, Centers for Disease Control and Prevention. 2015. Available from: www.cdc.gov/mmwr/preview/mmwrhtml/mm6426a3.htm?s_cid=mm6426a3_w

[21] HIV Among People Who Inject Drugs. Centers for Disease Control and Prevention, Centers for Disease Control and Prevention. 2021. Available from: www.cdc.gov/hiv/group/hiv-idu.html

[22] Fentanyl. Centers for Disease Control and Prevention, Centers for Disease Control and Prevention. 2021. Available from: www.cdc.gov/drugoverdose/opioids/fentanyl.html

[23] Prescription Behavior Surveillance System (PBSS). CDC National Center for Injury Prevention and Control, Centers for Disease Control and Prevention. 2021. Available from: www.cdc.gov/drugoverdose/pdf/pbss/PBSS-Report-072017.pdf

[24] Ongoing Emergencies & Disasters: Opioid Crisis. CMS, Centers for Medicare and Medicaid. 2020. Available from: www.cms.gov/About-CMS/Agency-Information/Emergency/EPRO/Current-Emergencies/Ongoing-emergencies

[25] Shihipar A. Declare racism a public health emergency. The New York Times. 2021. Available from: www.nytimes.com/2021/03/07/opinion/racism-public-health-emergency.html

[26] National Health Emergency Demonstration Grants. U.S. Department of Labor, Employment & Training Administration (ETA). Available from: www.dol.gov/agencies/eta/dislocated-workers/grants/health-emergency/phase-1-demonstration

[27] Rieder TN. We can't forget the nation's other epidemic. The Hastings Center. 2020. Available from: thehastingscenter.org/we cant forget-the-nations-other-epidemic/

[28] Increase in fatal drug overdoses across the United States driven by synthetic opioids before and during the COVID-19 pandemic. LA Country Public Health, Centers for Disease Control and Prevention. 2021. Available from: http://publichealth.lacounty.gov/lahan/alerts/CDC-HAN-00438%20Opioids12172020.pdf

[29] Goodnough A. Overdose deaths have surged during the pandemic, C.D.C. data shows. The New York Times. 2021. Available from: www.nytimes.com/2021/04/14/health/overdose-deaths-fentanyl-opiods-coronaviurs-pandemic.html

[30] Fuller T. San Francisco contends with a different sort of epidemic: Drug deaths. The New York Times. 2021, Available from: www.nytimes.com/2021/04/23/us/fentanyl-overdoses-san-francisco.html

[31] Wainwright JJ. Drug test results in US Before and After declaration of a national emergency concerning COVID-19. JAMA, JAMA Network. 2020. Available from: jamanetwork.com/article.aspx?doi=10.1001%2Fjama.2020.17694

[32] Signals Report COVID-19 Special Edition. Millennium Health LLC. 2020. Available from: www.millenniumhealth.com/news/signalsreportcovid/

[33] CDC COVID Data Tracker. Centers for Disease Control and Prevention, Centers for Disease Control and Prevention, Available from: covid.cdc.gov/covid-data-tracker/#datatracker-home

[34] Percy J. "Trapped by the 'Walmart of Heroin.'" The New York Times. 2018. Available from: www.nytimes.com/2018/10/10/magazine/kensington-heroin-opioid-philadelphia.html

[35] Opioid Overdoses. Philadelphia Department of Public Health, PDPH Health Information Portal. 2020. Available from: hip.phila.gov/EmergentHealthTopics/Opioids

[36] Whelan A. Fatal overdoses among Black Philadelphians Soared during the Pandemic, New Data Show. 2021. Available from: www.inquirer.com/news/overdoses-black-philadelphians-opioid-crisis-covid-19-20210422.html

[37] Whelan A. Even the pandemic doesn't slow down Philadelphia's drug markets. 2020. Available from: www.inquirer.com/health/opioid-addiction/coronavirus-pandemic-drug-supply-opioid-crisis-kensington-20200817.html

[38] Kerr T et al. Supervised injection facilities in Canada: Past, present, and future. Harm Reduction Journal, BioMed Central. 2017. Available from: harmreductionjournal.biomedcentral.com/articles/10.1186/s12954-017-0154-1

[39] Insite User Statistics. Vancouver Coastal Health, Vancouver Coastal Health, 2019. Available from: www.vch.ca/public-health/harm-reduction/supervised-consumption-sites/insite-user-statistics

[40] Ng J et al. "Does evidence support supervised injection sites?" National Center for Biotechnology Information (NCBI), Official Publication of The College of Family Physicians of Canada. 2017. Available from: www.ncbi.nlm.nih.gov/pmc/articles/PMC5685449/

[41] Pinkerton S.D. "Is Vancouver Canada's supervised injection facility cost-saving?" PubMed, U.S. National Library of Medicine. 2010. Available from: pubmed.ncbi.nlm.nih.gov/20653622/

[42] Andresen MA, Boyd N. A cost-benefit and cost-effectiveness analysis of Vancouver's supervised injection facility. PubMed, The International Journal on Drug Policy. 2010. Available from: pubmed.ncbi.nlm.nih.gov/19423324/

[43] Bayoumi AM, Zaric GS. The cost-effectiveness of Vancouver's supervised injection facility. National Center for Biotechnology Information (NCBI), Canadian Medical Association Journal. 2008. Available from: www.ncbi.nlm.nih.gov/pmc/articles/PMC2582765/

[44] Supervised Consumption Services. Drug Policy, Drug Policy Alliance. Available from: drugpolicy.org/issues/supervised-consumption-services

[45] Morgan H. CUES: The latest effort in the battle against the opioid crisis. City of Philadelphia, Department of

Public Health. 2018. Available from: www.phila.gov/2018-01-24-cues-the-latest-effort-in-the-battle-against-the-opioid-crisis/

[46] Clark P et al. Comprehensive user engagement sites (CUES) in Philadelphia: A constructive proposal. Phil Papers, Internet Journal of Public Health. 2018. Available from: philpapers.org/rec/LEECUE

[47] Larson S et al. Supervised consumption facilities—Review of the evidence. DBHIDS, Department of Behavioral Health and Intellectual Disability Services. 2017. Available from: dbhids.org/wp-content/uploads/2018/01/OTF_LarsonS_PHLReportOnSCF_Dec2017.pdf

[48] Kenney J, Krasner L. Criminalizing crack addiction was a mistake. Philly's approach to opioids must be different. The Philadelphia Inquirer, The Philadelphia Inquirer. 2018. Available from: www.inquirer.com/philly/opinion/commentary/opioid-safe-injection-sites-cues-comprehensive-user-engagement-philadelphia-jim-kenney-larry-krasner-20180215.html

[49] Gordon E. "What's next for 'Safe Injection' Sites in Philadelphia?" NPR, NPR. 2018. Available from: www.npr.org/sections/health-shots/2018/01/24/580255140/whats-next-for-safe-injection-sites-in-philadelphia

[50] United States v. Safehouse. SafehousePhilly, Safehouse. Available from: www.safehousephilly.org/us-v-safehouse

[51] Gabay M. The federal controlled substances act: Schedules and pharmacy registration. National Center for Biotechnology Information (NCBI), Hospital Pharmacy. 2013. Available from: www.ncbi.nlm.nih.gov/pmc/articles/PMC3839489/#:~:text=The%20Federal%20Comprehensive%20Drug%20Abuse, effective%20on%20May%201%2C%201971.&text=This%20federal%20law%20contains%203%20different%20titles

[52] U.S.C. Title 21 – Food and Drugs. GovInfo, U.S. Government Publishing Office. 2011. Available from: www.govinfo.gov/content/pkg/USCODE-2011-title21/html/USCODE-2011-title21-chap13-subchapI-partD-sec856.htm

[53] Roebuck J, Aubrey W. Judge: Philly supervised injection site proposal does not violate federal law. The Philadelphia Inquirer, The Philadelphia Inquirer. 2019. Available from: www.inquirer.com/health/opioid-addiction/safehouse-supervised-injection-site-ruling-philadelphia-mchugh-opioids-20191002.html

[54] Winberg M. Philly's safehouse to open nation's first supervised injection site after judge clearance. Billy Penn, Billy Penn. 2020. Available from: billypenn.com/2020/02/25/nothing-procedurally-improper-safehouse-injection-site-gets-judges-final-clearance/

[55] The United States Department of Justice, U.S. Department of Justice to Appeal District Court Ruling Regarding Drug Injection Sites. The United States Department of Justice, The United States Attorney's Office of the Eastern District of Pennsylvania. 2020. Available from: www.justice.gov/usao-edpa/pr/us-department-justice-appeal-district-court-ruling-regarding-drug-injection-sites

[56] Lieberman D et al. With overdose deaths up during the pandemic, Philadelphia fights for a legal safe injection site. NBCNews, NBCUniversal News Group. 2020. Available from: www.nbcnews.com/news/us-news/overdose-deaths-during-pandemic-philadelphia-fights-legal-safe-injection-site-n1235583

[57] Melamed S. Thousands are in Limbo as Philly Courts remain mostly stalled by coronavirus. The Philadelphia Inquirer, The Philadelphia Inquirer. 2020. Available from: www.inquirer.com/news/philadelphia-criminal-court-coronavirus-pandemic-shutdown-jury-trials-preliminary-hearings-20200801.html

[58] Jones A. Federal judge puts hold on philly safe-injection site. CourthouseNews, Courthouse News Service. 2020. Available from: www.courthousenews.com/federal-judge-puts-hold-on-philly-safe-injection-site/

[59] The United States Department of Justice. Appellate Court agrees with government that supervised injection sites are illegal under federal law; Reverses district court ruling. The United States Department of Justice, Office of Public Affairs. 2021. Available from: http://www.justice.gov/opa/pr/appellate-court-agrees-government-supervised-injection-sites-are-illegal-under-federal-law#:~:text=January%2013%2C%202021-Appellate%20Court%20Agrees%20with%20Government%20that%20Supervised%20Injection%20Sites%20are,room%E2%80%9D%20for%20illegal%20drug%20use

[60] Blanchard SK. Will Rhode Island be the first state to legalize safe consumption sites? FilterMag, Filter. 2021. Available from: filtermag.org/rhode-island-first-safe-consumption-sites/

[61] Blanchard SK. New Mexico the latest state to advance a safe consumption site bill. FilterMag, Filter. 2021. Available from: filtermag.org/new-mexico-safe-consumption/

[62] Blanchard SK. Biden quietly announces harm reduction among his drug priorities. FilterMag, Filter. 2021. Available from: filtermag.org/ondcp-biden-priorities/

[63] Kerr T et al. Impact of a medically supervised safer injection facility on community drug use patterns: A before and after Study. National Center for Biotechnology Information (NCBI), BMJ Publishing Group. 2006. Available from: www.ncbi.nlm.nih.gov/pmc/articles/PMC1352057/

[64] DeBeck K et al. Injection drug use cessation and use of North America's first medically supervised safer injecting facility. PubMed, Drug and Alcohol Dependence. 2011. Available from: pubmed.ncbi.nlm.nih.gov/20800976/

[65] Wood E et al. Impact of a medically supervised safer injecting facility on drug dealing and other drug-related crime. National Center for Biotechnology Information (NCBI), Substance Abuse Treatment, Prevention, and Policy. 2006. Available from: www.ncbi.nlm.nih.gov/pmc/articles/PMC1471778/

[66] Wood E et al. Changes in public order after the opening of a medically supervised safer injecting facility for illicit injection drug users. National Center for Biotechnology Information (NCBI), Canadian Medical Association Journal. 2004. Available from: www.ncbi.nlm.nih.gov/pmc/articles/PMC517857/

[67] Salmon AM. Five years on: What are the community perceptions of drug-related public amenity following the establishment of the Sydney medically supervised injecting centre? PubMed, International Journal of Drug Policy. 2007. Available from: pubmed.ncbi.nlm.nih.gov/17689343/

[68] Harris RE et al. Perceptions about supervised injection facilities among people who inject drugs in Philadelphia. UPenn Law, International Journal of Drug Policy. 2017. Available from:

Available from: www.law.upenn.edu/live/files/7788-perceptions-about-sif-among-people-who-inject

[69] Weekes J et al. Supervised Injection Facilities (SIFS). CCSA-CCLAT, Canadian Centre on Substance Abuse. 2005. Available from: www.ccsa.ca/sites/default/files/2019-04/ccsa-010657-2004.pdf

[70] Vital Strategies. Harm Reduction Receives Unprecedented $30 Million in Federal Funding through American Rescue Plan Act. Vital Strategies, Vital Strategies. 2021. Available from: www.vitalstrategies.org/harm-reduction-receives-unprecedented-30-million-in-federal-funding-through-american-rescue-plan-act/

[71] USCCB. Sharing catholic social teaching: Challenges and directions. USCCB, United States Conference of Catholic Bishops. 2011. Available from: www.usccb.org/resources/sharing-catholic-social-teaching-challenges-and-directions

[72] Byron WJ. The 10 building blocks of catholic social teaching. America Magazine, America: The Jesuit Review. 1998. Available from: www.americamagazine.org/faith/1998/10/31/10-building-blocks-catholic-social-teaching

[73] USCCB. Solidarity. USCCB, United States Conference of Catholic Bishops. Available from: www.usccb.org/beliefs-and-teachings/what-we-believe/catholic-social-teaching/solidarity

[74] National Harm Reduction Coalition. Principles of harm reduction. Harm Reduction, National Harm Reduction Coalition. 2014. Available from: harmreduction.org/about-us/principles-of-harm-reduction/

[75] Hawk M et al. Harm reduction principles for healthcare settings. Harm Reduction Journal, BioMed Central. 2017. Available from: harmreductionjournal.biomedcentral.com/articles/10.1186/s12954-017-0196-4

[76] Marlatt A, Witkiewitz K. Update on harm-reduction policy and intervention research. Annual Review of Clinical Psychology. 2010;**6**:591-606

[77] An opioid crisis foretold. The New York Times. 2018. Available from: www.nytimes.com/2018/04/21/opinion/an-opioid-crisis-foretold.html

[78] Feldman N. Safehouse debate in appeals court centers on two sentences in 'Crackhouse Statute.' WHYY, WHYY PBS. 2020. Available from: whyy.org/articles/safehouse-debate-in-appeals-court-centers-on-two-sentences-in-crackhouse-statute/

Chapter 2

Determination of Death: Ethical and Biomedical Update with International Consensus

Md. Shah Alam Panna

Abstract

Humanity has been confronted with the concept and criteria of death for millennia and the line between life and death sustains to be debated. The profound change caused by life support technology and transplantation continues to challenge our notions of life and death. Despite scientific progress in the previous few decades, there remain big variations in diagnosis criteria applied in each country. Death is a process involving cessation of physiological function and determination of death is the final event in that process. Legally, a patient could be declared dead due to lack of brain function, and still may have a heartbeat when on a mechanical ventilator. Though there is no point in supporting ventilation in a dead person, withdrawing a ventilator before the legal criteria for death may involve the physician in both civil and criminal proceedings. To identify the moment of death is vital to avoid the use of unnecessary medical intervention on a patient who has already died and to ensure the organ donation process, clear and transparent. The age-old standard of determination of death is somatic standard and cardiopulmonary standard. Harvard report (1968) defines *irreversible coma* as a replacement criterion for death and prescribed clinical criteria for the permanently nonfunctioning brain. The current unifying concept of death: *irreversible loss of the capacity for consciousness combined with irreversible loss of the capacity to breathe*. WHO (2014) adopted minimum determinant death criteria, acceptable for medical practice globally, achieving international consensus on clinical criteria to maintain public trust and promote ethical practices that respect fundamental rights of individuals and minimize philosophical and biomedical debate in human death. AAN (2019) endorses that the brain death is the irreversible loss of all functions of the entire brain and equivalent to circulatory death.

Keywords: definition of death, determination of death, bioethical issues of death, death determinant, controversies on death declaration, death declaration, international consensus on death declaration

1. Introduction

> *"The term brain death signifies that there is more than one variety of death. This is a serious misconception, perpetuated by such statements as "the patient declared brain dead at 3:00 a.m. on Thursday and died two days later." There is only one real phenomenon of death that clinicians and families struggle to recognize."*

IntechOpen

> *"A case of Elaine Esposito who lapsed into coma following surgery on August 6, 1941 and died on November 25, 1978, 37 years and 111 days later."*
>
> *President council (2008) of bioethics**

Death is often considered in terms of medical, legal, ethical, philosophical, societal, cultural, and religious rationales. The biomedical definition of death is primarily a scientific issue supported by the best available evidence. A medical practitioner has certain ethical and legal responsibilities regarding death, such as the effort for prevention of death, determination of death, determination of time/moment of death, declaration of death, issuing the certificate, and if needed, autopsy or organ removal for transplantation. That aspect has a lot of ethical, legal, emotional, and scientific issues. Dying is considered as a process, which affects different functions and cells of the body at different rates of decay. Doctors must decide at what moment along this process there is permanence and death can be appropriately declared. Diagnosis of death and a record of the time of the death, in most countries, are the legal responsibility of a medical practitioner. Determining the moment of death is vital to avoid the use of unnecessary medical interventions on patients who have already died and to make sure that the method of organ donation is obvious and transparent. Also, the time of death is important because of survivorship clauses in wills.

For the millennia, human has struggled with the concept and criteria of death, and thus, the line between life and death continues to be debated. The profound changes caused by the life support in organ failure, organ substitution technology, and transplantation still continue to challenge our notions of life and death [1]. Despite scientific progress in the last few decades, there remain big variations in the diagnosis criteria applied in each country with legal regulations resulting in misunderstandings among the public and health care professionals. Since the ample decades, the academic literature and the media have raised the voice in alarming language in issues of death determination and dead donation practices [2]. Difficulty arises to distinguish valid scientific critique from those criticisms supported by the fear of death itself, mistaken diagnosis or a premature declaration of death, or the fear of retrieving organs from the living.

The challenges in discussions about death are complex due to philosophical, religious, and cultural differences in the concept and definitions of death; debate about ethics, law, and religion; problems in performing research and the resultant shortfall in information and evidence on various aspects of the dying process; dispute in the validity of death determination practices; lack of understanding and/or awareness by general public and health professionals; last but not the least the emotionally charged nature of the subject matter. There are plentiful ways of dying but just one way to be dead. Hence, the baseline determination of death criteria should be rigorous, global, and acceptable for medical practice worldwide, while remaining respectful of diversities. International consensus on the clinical criteria for the death determination is of central importance to preserve public trust and promote ethical practices that respect the fundamental rights of people and promote quality health services [3].

In medical practice and law, the separation between being alive and dead should not be ambiguous. It designates the moments that follow events such as no medical or legal need to maintain resuscitation or life support, loss of personhood and individual rights, decedent's legal will execution, disposal of the estate, life insurance

* McWhirter N. The Guinness Book of World Records. New York: Bantam Books; 1981. p. 42 [citing the case of Elaine Esposito who lapsed into coma following surgery on August 6, 1941 and died on November 25, 1978, 37 years and 111 days later]

settlement, burial or cremation of the body for final disposition, and religious or social ceremonies to mark the end of a life [4]. Dying is not an event rather a process, which affects various functions of the body at different rates of decay. The physician must confirm the moment along this process that there is permanence and death can be accurately declared [5]. Biological criteria of death are associated with biological features and irreversible loss of certain cognitive capabilities [6]. A patient could be declared dead legally as lack of brain function and may still have a heartbeat when on a mechanical ventilator. Though there is no justification in supporting the ventilation of a dead person, withdrawing the ventilator before the legal criteria of death may involve the doctor in both civil and criminal proceedings. The legitimate moment of death could be a wide range of time after the death has actually occurred. Many accident victims actually died at the scene of the accident but were declared dead officially on arrival at a hospital.

The scientific, biological, and medical aspects of the determination of death are still controversial. Certain ancillary and/or complementary laboratory tests could also be useful in situations where clinical testing cannot be executed or if confounding or special conditions are present. It had been recognized that there are limitations to the utilization of a number of these tests and further work will confirm the reliability of these tests. Death is a biological phenomenon, with profound social, religious, and psychological traditions, but very little background experience and available scientific information. The understanding of the biological aspect has gradually developed and strengthened as a direct result of technology, cell biology, organ donation, and transplantation, but was inadequately adjust in law, health policy, and bioethical discourse. Organ donation has forced the understanding of moment of death and acceptance or persisting controversy of where that line is.

It is urgent time demanding notion to adopt a minimum determination of death criteria to be acceptable for medical practice worldwide to achieve international consensus on clinical criteria to maintain public trust and promote ethical practices.

2. Philosophical, religious, and bioethical discourse/debate

The concepts and practices of death undoubtedly are influenced by values and social practice. The definition of death affects not only that consider to count as death, but also questions of grieving, medical treatment, asset disposal, organ donation, and a myriad of other legal and ethical issues [7].

The philosophical investigation of human death has focused on some overarching questions—What is human death? The conceptualization (definition) of death is the answer to this ontological question that defines death as the irreversible cessation of organismic functioning along with the irreversible loss of personhood. Next question, how can be determine that death has occurred? The answer is epistemological one, which furnishes both the *general standard* (criterion) for determining that death has occurred and *specific clinical tests* to show whether the standard has been met in a given case. Examples are traditional cardiopulmonary standard or neurological standard [8].

Finally, how do the deaths relate, conceptually, to the essence and identity as human persons? The metaphysics of the body and soul does so in terms of the logical dualism between the material body and consciousness or the immaterial mind. In the philosophy of mind, mental phenomena are nonphysical and thus distinct and separable from the body. The dualism of body and soul/mind suggests that while being a person is, undoubtedly, a matter of having a biologically human body. The existence of a person entails the presence of a thinking being, which has reason and reflection, and can consider itself as itself, in different times and places. The individual identity of psychological persons is dependent on the brains' neurophysiology [9]. The brain death need not be considered as biological death rather a proxy for the loss of individual identity, that is, personhood [10]. When a person has died, it does not merely mean that some biological entity no longer functions. It means some unique mind or person, realized as a cognitive or psychological entity, has ceased to exist. The personhood admits of application of the terms life and death. It has been exceedingly rare for the demise of a biological human organism to take place sometime after the death of a person. Artificial life support can maintain the biological life of an individual in the absence of their continued psychological existence. Such brain dead individuals have been considered living cadavers and twice dead [11]. Human life is operationally defined by the onset and cessation of organismal function [12]. There are two different meanings to human death being alive and having a life, the notion of personhood allows us to focus on the autobiographical meaning of death—the loss of a person [13].

Other philosophical questions—When does a human being die? Is the organismic and denouement conception of death have any practical use? Schofield et al. present a definition of death focused on the final denouement of human beings as biological organisms. According to their view, the moment of death is the last process in bodily functions that maintain homeostasis and finally ceases [14]. Reducing death to the biological denies an important characteristic of being human, intellectual, or psychological nature. The conception of death acknowledges the cognitive aspect of human existence, at the same time, accommodating embodiment, that we both have and are biological bodies [15].

Generally, people believe that death terminates the whole existence of a person. According to Christians' belief, death puts an end to human existence on earth but does not end existence, instead opens an entrance into another sphere where existence continues either in heaven or hell after the final judgment that everybody will face it. Death is considered *a type of sleep* as the sleeper does not cease to exist while his body sleeps. Therefore, a dead person continues to exist despite the absence from the region in which those who remain can communicate with him. Sleep is understood to be temporary, but the unconsciousness in the dead is seen to be permanent [16]. Muslims and Jews have the same belief. The Quranic verses distinguish life and death, as sleep is a form of reversible unconsciousness (life) comparable to death as irreversible unconsciousness, both are the commandment of Almighty creator of the universe. With the moment of this commandment, the total integrating and coordinating mechanism of the human body is irreversibly lost and the person has no relation with this world. Finally, all religious beliefs support that death is the separation of the soul from the body. Plato defines death, as nothing else but the separation from each other of two things, soul and body [17].

Life is fundamentally grounded on the continuation of individual and collective cell function, dependent on the supply of nutrients and oxygen. Cell biology has exhibited that a layer of human cells, separated from the human organism, could also be grown in laboratory culture pending till bathed in a steady supply of nutrients and oxygen. The human being, a complex package of trillions of cells organized into organ systems, requires a cardiopulmonary delivery system for oxygen and nutrients to reach the cells. The development and evolution of modern cardiopulmonary resuscitation evolving into cardiopulmonary support technologies have been important advances informing our concepts of life and death.

The introduction of advanced medical technology poses new problems for the old standards that constitute death. The values automatically shape thinking of the death of a person, not merely a descriptive, scientific concept, but unequivocally contain evaluative content. The changing frontiers of the death drive to confront basic questions of persons and values that will adapt to address future questions.

It is vital to examine the evaluative content of concepts and practices relating to death, and reflects on what it is that we value or should value in persons. The philosophical definitions of death in the absence of indisputable objective signs of death should be considered, loss of integrative functioning of the whole organism, failure to engage the environment spontaneously by respiration, loss of consciousness and sentiency, and the separation of some vital principles from the body.

The neurological criteria for death represent an interesting advance in the ways of responding to changes in death and dying. The development of medical technology and life support techniques insist increasingly on precise notions to identify the most important aspect of neurological lives. However, the whole brain standard of death suffices in the vast majority of cases, but does not fully line up the value in persons. Time has come to decide the position of the current brain death standard as it mismatch with the values and negative consequences in determining death and in organ donation. Advances in technologies seem as if they will inevitably make this question inescapable. The prominence of ***biomedical criteria*** relying on brain death reduces the impact of metaphysical, anthropological, psychosocial, cultural, religious, and legal aspects disclosing the real value and essence of human life [6]. Should we retain the current brain death standard despite its mismatch with our values and despite negative consequences in determining death and in organ donation? Is a human being with total brain failure dead?

3. Definition of death

Ideally, the definition of death would link the concept of life or death with its clinical manifestations as closely as possible that fall in both two categories, the *philosophical* (conceptual), the understanding of essential differences between life and death, and *empirical*, which is to determine clinical signs, tests, or criteria that separate life and death most accurately.

Death is the transition from being a living mortal organism to being something that, though dead, retains a physical continuity with the once-living organism. Death is a process involving the cessation of physiological functions and the determination of death is the final event in that process. Death is a gradual process at the cellular level with tissues varying in their ability to withstand deprivation of oxygen. A distinction is now being made between death at the cellular and tissue levels and death of the person. Sydney declaration states, clinically, death lies not in the preservation of isolated cells but in the fate of a person. Korein's view of the life of the multicellular organism as a whole could no longer be explained in terms of a cellular task alone. The life of a typical unicellular organism encompasses fundamental tasks of the metabolic and reproductive attributes of a particular organism, empowering it to amplify in a direction of decreased entropy production (bacteria, amoeba, or zygote). In a multicellular organism, a large mass of cells could be alive but this does not indicate that the organism as a whole was alive. Machado refused the hypothesis that an explanation of death should include the function that contributes to the key human attributes and the highest level of control in the hierarchy of integrating functions within the human organism [18–24].

The full version of death includes three unique ingredients such as the definition of death, yardstick of brain death, and the tests to prove that the standard has been satisfied. The definition of death is typically a philosophical task, while the criteria and tests are medical tasks. Particular standards and tests must match with a given definition. The definition must represent attributes that are so important and significant to a living entity that its absence is designate death [25, 26]. The nonfunctioning entire brain provokes the permanent cessation of the functioning of the organism as a whole.

Biologically death is defined as the extinction of biological properties of life. Human death can be defined as the irreversible cessation of three interdependent and interlink vital functions of the body—the tripod of life (heart, lung, and brain). Another way death can be defined as a person is said to be dead, if he cannot take up spontaneous respiration or maintain circulation. There is growing medical consensus in a unifying concept of human death, which involves the irreversible loss of the capacity for consciousness, combined with the irreversible loss of the capacity to breathe.

Uniform determination of death (UDDA) act defines death as, an individual who has sustained either irreversible cessation of circulatory and respiratory functions, or irreversible cessation of all the functions of the entire brain, including the brain stem, is dead. Montreal forum defines death as the irreversible loss of the capability for consciousness and loss of all brainstem functions. That could result as a consequence of permanent stoppage of circulation and/or after catastrophic brain injury. In the determination of death, "permanent" refers to the cessation of function that cannot resume automatically even not be restored through intervention. The determination of death must be made in accordance with accepted medical standards.

4. Pathophysiology in death process: brain failure and ventilator support

The presence of the two vital functions, circulation and respiration in a body, is a sure sign of life. The patient who was diagnosed with entire brain failure and has been pronounced dead the vital functions are dependent on external support from the ventilator. The supporter of neurological standard designates these apparent signs of life are artifacts of the mechanical support that conceal the very fact that death has already occurred. To judge that logic, the essential facts of mechanical assistance for these vital functions be achieved if the interrelationship of three-body systems involved in breathing and circulation is understood. The three systems are the heart and circulatory system, the lungs and respiratory system, and the central nervous system. The pathophysiological processes that eventually end in mortal condition, *total brain failure* engage not only the central nervous system but also the circulatory and respiratory systems.

The prime functions of respiration are ventilation and diffusion. The ventilation involves both inhalation and exhalation; the diffusion involves the exchange of oxygen and carbon dioxide between atmospheric air and blood. The respiratory system brings atmospheric air by inhaling process to the alveoli where oxygen from the atmospheric air is able to move into the blood by the process of diffusion. The exhaling process of breathing facilitates to rid the body of the waste products—carbon dioxide. The walls of the alveoli are extremely thin, formed to facilitate the diffusion of gases between the sacs and the blood vessels. Oxygen is essential to the continued metabolic work of the trillions of cells in the body. The absence of an endless delivery of oxygen, brought into the body through inhalations and transported to the tissues by the *circulatory* system, the cells, tissues, and organs of the body would stop functioning. The gas exchange is facilitated by the contraction and the relaxation of the muscles of respiration and the diffusion of gases into the blood across the lining of the tiny alveoli.

The CNS plays a crucial role in maintaining an organism's vital functions. The reticular activating system of the brainstem is also critical to the organism's conscious life, essential for maintaining a state of wakefulness, which is a prerequisite for any of the activities associated with consciousness. The contraction of the muscles of respiration is brought about by a signal sent from the respiratory center located at the brainstem. A relatively high level of CO_2 in the blood stimulates the respiratory center to send a signal to the muscles of respiration, which excites them to contract. For life to continue, the CO_2 must be expelled and new oxygen brought in. Other parts of the CNS also be involved in signaling the muscles of respiration to contract, like *conscious breathing* where a person deliberately controls the depth and pace of breathing or *without conscious effort* as during physical exercise. If the respiratory centers of the brainstem are disabled, the organism will not make any respiratory effort. The chest will remain absolutely immobile and the body's need for oxygen will go unanswered [27].

To prevent the death of the organism, some external device (mechanical ventilator) for the breathing process is essential. The mechanical ventilator works by altering the pressure in the lung cavities in order that oxygen-rich atmospheric air will travel down and CO_2-rich air will travel back up the respiratory tree. Gas exchange in the lungs will be of no benefit to the patient unless the blood is kept moving as well. Incoming oxygen must be delivered to tissues that required it, and accumulating carbon dioxide must be a shift to the lungs for expulsion from the body. Hence, a ventilator will help the patient as long as another vital system is functional, constituting the heart (working as a pump) and network of arteries, veins, and capillaries. The movement of blood occurs only within the body, whereas the movement of air is an exchange between the body and the surrounding atmosphere. Another relevant rationale of external support of vital systems is the indisputable fact that there is no part of the CNS that is absolutely essential for heart contractions within the way as the respiratory center in the brainstem is unconditionally essential in breathing. The heart is the most essential active part of the circulatory system and the vessels of circulation, being rigid plumbing lines that passively convey blood, pumped by the heart, are living tissues that undergo changes (some driven by CNS) to sustain a proper blood pressure. Patients of ventilator support must also be given drugs to maintain the blood pressure in a healthy range. Ventilator support designates the external supports of vital functions of breathing and circulation, in lieu of breathing effort of organism, stimulated by the respiratory centers of CNS, an external device moves the lungs and facilitates the inflow and outflow of needed air. It offers the heart muscle still to function, as the myocardium, like other cells in the body, needs oxygen to stay alive. The argument for the neurological standard of determination of death begins with facts that the respiratory motion supported in this way is not in itself a symbol of life, rather an artifact of technological intervention. Neither a beating heart, in this instance, a symbol of life, or merely the continuation of a spontaneous process would quickly cease if the ventilator is withdrawn [27].

Loss of the ability to breathe is not a sufficient condition for declaring that an individual has died, along with other functions indicative of life must be lost and functional losses must be irreversible. The inability to breathe automatically is insufficient for pronouncing death can often easily be dispelled by considering neurological injury that deprives a patient of the power to breathe and yet leave untouched the ability to continue activities dependent on other parts of the CNS. Patients with high spinal cord injuries remain awake and alert but dependent upon ventilators for respiratory support. Moreover, deprivation of all functions of the CNS is not a sufficient criterion for declaring death if this stoppage of function is not permanent. There are critical care cases that reveal the significance of this criterion such as a patient in a deep, nonbreathing (apneic) coma during a critical emergency and therefore assist in ventilator allowing time for CNS functions to return. Sometimes a full recovery of CNS functions happens, though the functions that return will only be enough to abandon the patient in a vegetative state and will be labeled as a persistent vegetative state PVS.[†] The deep, nonbreathing coma that the patient was in prior to waking into the vegetative state could not have been dead since the loss of functions proved to be reversible.

5. Historical landmark in biomedical aspect of death determination

5.1 Medieval landmark: transition from heart to brain

Humans have long used criteria and technology to assist in the diagnosis of death. The link between breath and life is equally as ancient and found in both Genesis (2:7) and the Qur'an (32:9). Somatic criteria, such as the presence of decomposition and rigor mortis, are the oldest in human history. Over 800 years ago, when Maimonides codified the diagnosis of death as the absence of the heartbeat and respiration with cooling of the body [28], he was likely documenting a standard used from down of civilization.

In the eighteenth century, the physician was confirmed about death if the heart and lungs break off, but lacked adequate tests to certify it. In the twentieth century, the moment of death became less clear, and thus, the tests physicians had finally perfected proved insufficient. Historically, until the early twentieth century, physicians' inexperience in human anatomy and physiology left them poorly equipped to accurately test for death. From the eighteenth through the mid-twentieth centuries, a person was declared dead when the heart stopped beating and lungs ceased to function. In the early part of the twentieth century, while the standard to check death was well established, the understanding of when the death occurred became the subject of great debate. The fear of premature burial was replaced by the fear of apparent death sustained by life support systems. These issues reach a climax in the latter part of the twentieth century when the cardiorespiratory definition of death was reevaluated and a novice addition of brain death was introduced. Intensifying new questions as to the moment of death, the brain death criterion demands further revision of the empirical tests. The nature of death, however, does not lend itself to one discipline rather considers metaphysics, sociology, theology, and medicine. Historically, the irreversible stoppage of heart and lung functions constituted death as the absence of heart and lung activity immediately leading to failure of the entire organism. It has become apparent that cardiac and respiratory activities were significant for separating the living from the dead. The moment of death was firmly estimated but the task of confirming criteria to check for irreversible quiescence of functions proved more challenging and often had catastrophic consequences. A consensus emerged that once the heart and lungs ceased to function the person was dead, although the empirical criteria to test for death were suspect. Because of this critical divide between theory and practice, instances of premature burial occurred. To safeguard premature burial date back to antiquity with the Thracians, Romans, and Greeks, each waited 3 days for putrefaction to start before burying their dead. The Romans took a more extreme approach by amputation of a finger to ascertain if the stump bled, in addition to calling out the person's name three times while on the funeral pyre. Hence, the premature burial was a great worry, though it did not attain climax until the eighteenth century, accelerated by the intellectual climate. The knowledge and scientific revolution instituted a radical change in the insight of life and death [29].

[†] Jennett B, Plum F. Persistent vegetative state after brain damage. A syndrome in search of a name. Lancet. 1972;1(7753):734-737

Belief in the afterlife was not as important as life here due to the works of Bacon, Descartes, and Galileo, which emphasized the notion that life might be improved if not perfected by scientific manipulation. There is little practical obligation to worry oneself with an afterlife if this life could be manipulated by the art of medicine. Revulsion (drawing of disease) by the dissection of cadaver found in the sixteenth and seventeenth centuries as the study of human anatomy revealed the secrets of the *belle mécanique*, or the beautiful machine [30]. Fears of premature burial appear to have culminated in the eighteenth century, when George Washington made his dying request and Jean-Jacques Winslow in 1740 famously stated that putrefaction is the only sure sign of death. In 1833, Dungson also voiced commencement of putrefaction as a certain sign of real death [31]. Traditionally, the physician uses the basic cardiopulmonary standards as heart or lung functioning criteria to determine the death. The physicians palpate pulse, listen for breathing, hold a mirror in front of the nose to test for condensation and look if the pupils were fixed. William Harvey, in seventeenth century, first described the circulation of blood and the function of the heart and under this concept, death was when the heart and circulation have stopped. Ibn al-Nafīs (died 1288), an Arab physician, wrote about pulmonary circulation 300 years before it was discovered in Europe [32–35].

During this era, fear for early burial was so prominent that led to the establishment of waiting mortuaries and security coffins with alarm mechanisms and permanent air supply. The "Academy of sciences prize" was awarded in 1846 to Dr. Eugene Bouchut for his best work on the "signs of death and the means of avoiding premature burials." He suggested the utilization of the stethoscope, invented in 1819 by Laennec, as a technological aid to diagnose death. Other popular practices for death determination were inserting leeches near the anus, applying specially designed pincers to the nipples, or piercing the heart with a long needle with a flag at the end, which wave if the heart is still beating. Bouchut suggested that a person could be declared dead if a heartbeat was absent for 2 min. He extended the period to 5 min, in the face of opposition [36–39]. Case reports from physicians (Harvey Cushing) writing around the beginning of the twentieth century had evident that patients of cerebral pathology would die from respiratory arrest and subsequent circulatory collapse. Loss of electrical activity in the brain and cerebral circulatory arrest might signify human death that was evident in subsequent decades. The advent of mechanical ventilation, halting the inevitable circulatory collapse that follows the cessation of spontaneous respiration with the advent of mechanical ventilation, and the relevance to diagnosing death using neurological standard were understood.

In 1959, two historical landmarks were published, Mollaret and Goulon proposed the term *coma de′passe′* for an irreversible state of coma and apnea, and also, Pierre Wertheimer's group, a few months earlier, proposed neurological standard for death determination, that is, *death of the nervous system* [40, 41]. Those standards are practiced widely as an indicator of medical futility and a point at which ventilation might be stopped. Using neurological criteria, Belgian surgeon Guy Alexandre carried out the first transplantation from a heart-beating donor in 1963 and Christiana Barnard performed the first heart transplantation in 1967, after DCD who satisfied the criteria for coma de′passe′ [40].

5.2 Papal allocution (1957): prolongation of life

A group of anesthesiologists observed problems of sustaining the body alive in the absence of total brain function. This problem was presented to Pope Pius XII and resulted in the publication of a papal allocution describing that *the death* declaration *was not the province of the church*. Acknowledged, *it remains for the doctor to* offer *a* transparent *and precise definition of death and* therefore, *the moment of death*. Another important point in the relation of the "prolongation of life" was that death should not be opposed by extraordinary means in hopeless conditions, though precise phenomena of hopelessness and extraordinary were not stated. Thus, in such hopeless cases resuscitation could be discontinued and death be unopposed [42].

The papal allocution culminate research, by three categories of French neurologists and neurophysiologists during 1959, separately studied comatose and apneic patients separately, narrated terms death of the "systema nervosum and coma de´passe´" translated as beyond coma or ultra-coma and subsequently by others as irreversible coma. These patients were respirator dependent, in an unresponsive coma, and areflexive. EEG and deep intracranial electrical activity were entirely absent. The investigators' conclusion was that the brains of these patients were irreversibly dysfunctional. The WMA ethical committee and its council undertake dialogue and conference on death, 2 years earlier the first heart transplant by Christian Barnard in 1967. Wijdicks wrote that the first idea for the formation of the Harvard committee was recorded in a letter from Henry Beecher to Robert H. Ebert in September 1967. The Sydney and Harvard committees worked in parallel for several months, without either being aware of the other's work [41–45].

5.3 Harvard Ad Hoc committee report and Sydney declaration: new definition

The year 1968 was a crucial time for defining human death on the neurological ground and a milestone event in the history of medical science. On August 5, 1968, the Ad Hoc committee of the Harvard medical school to examine the definition of brain death published a report, as ***irreversible coma***, in the JAMA [46]. On the same day, the 22nd World Medical Assembly, meeting announced the Declaration of Sydney [47–49], a pronouncement on death that is less often quoted because it was overshadowed by the impact of the Harvard report. The delegates from 26 countries of 64 WMA member nations met in Sydney, Australia, to hold the 22nd assembly. The WMA had been worried about a new definition of death, to formulate a report of death under the new circumstances in an epoch of advances in resuscitation, and the increasing need to find organs for transplantation. The concept of brain death was formulated in this landmark report as irreversible coma. Though brain death has been widely accepted for the determination of death globally, many controversies yet to be settled. The concept evolved as a result of the convergence of several parallel developments including advances in resuscitation and critical care, research into underlying physiology of consciousness, medical futility, and ethics in end-life-care.

Since 1968, the concept of brain death has been extensively analyzed, debated, and reworked. Still, there remain much misunderstanding and confusion, especially for the general public [50]. The Declaration of Sydney touched on key philosophical issues on human death. It proclaimed that in most situations physicians could diagnose death by the classical cardiorespiratory criteria. In spite of this, two modern practices in medicine force them to revise the time of death: first the ability to maintain circulation by artificial means and second the use of cadaver organs for transplantation. The essential public addresses death as a progressive process at the cellular level with tissues varying in their capability to cope with deprivation of oxygen, but clinically death "lies not in the preservation of isolated cells but in the fate of a person." Also, it is described that the death determination must be grounded on clinical judgment, supplemented if necessary by a number of diagnostic aids, emphasizing the EEG. Nonetheless, it asserted that the overall judgment of the physician could not be replaced by any ancillary test. The declaration went further, proposing a more philosophical and conceptual explanation about the relationship between death and the fate of a person. The Harvard committee did not provide a clear concept of death but emphasized a clinical explanation of brain death, describing in detail the anatomical substratum and tests. The Sydney declaration did not use the term brain death but declared the clinical judgment for death determination and the Harvard committee, although mentioned the term brain death, finally select irreversible coma along with a detailed set of clinical criteria for death declaration. Both the Sydney and the Harvard committees suggest the use of EEG. For the purpose of the death diagnosis and transplantation, the Sydney declaration advocates two or more physicians not involved in transplantation should make the diagnosis, while the Harvard committee voiced that the death declaration should be made first, and then, physicians not involved in the transplantation procedure should be the one to turn off the respirator. Both committees justify a legal regulation of this issue [49]. Sydney declaration was amended at 35th WMA, by the addition of a key point declaring that "It is essential to determine the irreversible cessation of all functions of the whole brain, including the brain stem" for diagnosis of brain death but the EEG was not mentioned and no other issues were modified [49, 50].

5.4 President's commission report on medical, legal, and ethical issues in the determination of death: definition of death and UDDA

In July 1981, the President commission for the study of ethical problems in medicine and behavioral research published a report, *defining death*, to the President, Congress, and the relevant US government departments. It proclaimed that a person is dead, who has experienced either *irreversible stoppage of circulatory and respiratory functions*, or *irreversible cessation of all functions of the whole brain, including brain stem*. The death determination must be practiced in accordance with accepted medical standards. President's Commission report permitted consolidation of the whole-brain criterion of death [51].

A scientific basis was suggested to justify brain death with the theory of the brain as the central integrator of the body. According to this theory, the organism becomes a rapidly disintegrating collection of organs following the brain death (BD). Consequently, the concept of BD is not only an ethical and/or social concept or a matter of values, rather a matter of scientific facts such as irreversible stop page of functioning of the organism as a whole is death. The guiding principles of irreversible cessations of functioning of the entire brain are absolutely correlated with the permanent cessation of functioning of the organism as a whole as the brain is necessary for the functioning of the organism. The brain integrates, generates, interrelates, and controls complex bodily activities. A patient on a ventilator with entirely destroyed brain is merely a group of artificially sustained subsystems since the organism as a whole has ceased to function. President's Commission report also supports that rationale, convincing the gravity of the brain and recognized the profound instability of the brain-dead organism. In adults who have an irreversible stoppage of the whole brain's function, the mechanically generated functioning could exist only for a limited time as the heart usually stops beating within 2–10 days [51].

The enabling legislation for the President's Commission directs it to study the ethical and legal implications of the matter of defining death, including the probability of developing a uniform definition of death [51]. The central conclusions were that the recent developments in medical treatment necessitate a restatement of the standards traditionally recognized for determining that death has occurred and such a restatement ought preferably to be a matter of statutory law, which should be uniform among all the states. The definition embodied in the statute ought to address general physiological standards instead of medical criteria and tests, which will change with advances in biomedical knowledge and refinements in technique. The death is a unique episode that could accurately be confirmed either on the traditional grounds of permanent cessation of heart and lung functions or on the basis of permanent loss of functions of the entire brain. Any statutory definition must be separate and distinct from provisions governing the donation of cadaver organs and any legal rules on decisions to terminate life-sustaining treatment. American Bar Association, American Medical Association, and the National conference of commissioners on uniform state laws together have declared the statute, the Uniform Determination of Death Act (UDDA) affirmed: "an individual is dead who has **sustained** either, *the irreversible cessation of circulatory and respiratory functions*, or *irreversible cessation of all functions of the entire brain, including the brain stem*." A determination of death must be made in accordance with the accepted medical standards [51].

The UDDA is a statute, to address the societal problem created in the mid-twentieth century, due to the development of mechanical ventilation and other organ-sustaining technologies, to support permanently brain-injured individuals. The justification of the UDDA was to establish a uniform definition of death, determined by *acceptable medical standards*, that was transparent and socially accepted. The President's Commission and the UDDA considered death to be a unique episode in spite of causation, resulting from either irreversible failure of the brain or circulatory function. The act acknowledged the biological facts of universal applicability while seeking to safeguard patients against ill-advised idiosyncratic pronouncements of death. The UDDA viewpoint was supported by the majority of medical and legal authorities, the original UDDA wording also supported by the AAN. The neurologic insults may cause temporary stoppage of multiple brain functions, leading to disorders of consciousness, and the irreversibility component for the brain death criteria requires that these functions have ceased permanently, with no hope of resumption through clinical intervention or auto-resuscitation. Several medical associations support the UDDA definition of death and have participated in guideline development pertaining to the establishment of brain death in adults and children. A patient declared dead legally by considering neurologic criteria in all the state of USA except the state of New Jersey, however, allows for religious exemptions to the declaration of brain death if family members object. In such cases, death is not declared until the patient has met cardiopulmonary criteria for death [52–55].

5.5 Task force recommendations (1987): the guidelines for the diagnosis of brain death in infants and children (pediatrics guideline)

In the executive summary update of task force recommendations, declare requisite for the diagnosis of brain death in children of two neurologic examinations is performed by two independent physicians and two apnea tests, both of which may be organized by the physician managing ventilator care [56, 57]. Examinations should follow an observation period of 24 hours for neonates less than 30 days old and 12 hours for older infants and children up to age 18. It is significant to note that there may be institutional variance in the way these criteria are interpreted, and pediatricians may adapt their brain death testing methods to take into account the age-related anatomical and physiological differences between neonates, infants, and children. Parents and other family members of children undergoing brain death testing may require close attention and additional support [58, 59]. The pediatric guidelines were updated in 2011 by the American Academy of Pediatrics. A recent study reveals widespread disparities in adherence to the guidelines nationwide. It is essential to follow a standardized process to ensure accuracy in the diagnosis and inconsistencies in diagnosis could lead to false-positive brain death determinations, which could erode the public trust in the ability of physicians to declare death [58–60].

5.6 White paper on controversies of determination of death by president council of bioethics: total brain failure (2008)

In December 2008, the President's council on bioethics published a white paper (controversies in the determination of death) in which the neurological standard was carefully reexamined [27]. The council built the insight in biological reality by appropriately describing the clinical and pathophysiological understanding of brain death, which offers substantial reassurance to the ultimate validity of the neurological standard. It effectively gives a new foundation to the justification for the neurological standard of death. The council strongly agreed that "*Relaxation of the DDR is a morally unacceptable and logically specious way to deal with the uncertainties of the criteria for the death of the donor.*"

The council works was a historical decision that answers lot of *controversies and philosophical debate* in light of sound biological and pathophysiological evidence; debate on several controversies in the determination of death includes those arising in the context of controlled DCD with a primary focus on the clinical and ethical validity of neurological standard; controlled DCD is analyzed and the traditional cardiopulmonary criteria, also voiced concerns about the difficulty of safeguarding adequate end-of-life care for the patient-donor. The principal argument was—Does a diagnosis of *whole brain death* mean that a human being is dead? In other words, does the neurological standard rest on a sound biological and philosophical basis? [27].

Whether a patient in the condition of total brain failure is actually dead and can it be said with sufficient certainty to ground a course of action as the mortal remains of a human being. To ascertain those, up to this time, two facts about the diagnosis of total brain failure have been taken to provide basic support for a declaration of death: first, that the body of a patient with *total brain failure* diagnosis is no longer a *somatically consolidated whole*, and, second, that the capacity of the patient to sustain *circulation* will cease within a definite span of time. Another dispute, a patient with total brain failure is no longer able to carry out the basic work of a living organism. The patient has lost permanently the openness to the surrounding environment and the ability and drive to act on this environment on his own behalf. The respiratory function and cell metabolism sustained by mechanical ventilation are not due to *spontaneous* respiration. The council on bioethics acknowledges that such interventions with mechanical ventilation may preserve certain integrative bodily functions in patients declared dead by neurologic criteria, and such integration is not enough to define these patients as living. Patients who experience the neurologic criteria for brain death can no longer conduct the definitive work of a living organism, which is to be receptive to and act upon its environment in order to acquire the needs to preserve itself, such as breathing spontaneously, withdrawing from pain, or sleeping and waking. While such behaviors do not justify self-consciousness, they verify that the organism is alive. However, the patients kept alive *artificially*, by pacemakers, defibrillators, vasopressors, ventricular assist devices, artificial nutrition, and hydration, are not, by that fact alone, considered to be *dead*. A living organism participates in self-sustaining, need-driven activities essential to and constitutive of its trading with the surrounding world. These activities are genuine signs of active and existing life. A judgment that the organism as a whole has died can be made with confidence if these signs are lost and the activities had stopped.

Another view of the neurological standard was also pointed within the council for certainty about the vital status of patients with total brain failure, the only rational and defensible conclusion of such patients are severely injured, but not yet dead. Hence, only the traditional signs of permanent cessation of heart and lung function should be used to declare a patient dead. Accordingly, medical interventions for patients with total brain failure should be withdrawn only after they have been judged to be *futile*, in the sense of medically ineffective and non-beneficial to patients and disproportionately burdensome. The judgment must be done on the basis of ethical grounds considering the whole aspect of the particular patient and not merely the biological facts of the patient's condition. Then, the interventions can and should be withdrawn so that the natural course of the patient's injury can reach its inevitable terminus. Preparation for burial or for organ procurement is morally valid only when medical intervention has been judged as futile and the heart of the patient has stopped beating [27].

The understanding of medical futility [61, 62] has been developed in several papers by Edmund D. Pellegrino. Futility is the condition of a patient's disease, which is beyond medical rescue, such as beyond the powers of medical technology to help. Clinical futility is present when any medical intervention is considered as ineffective, non-beneficial, and disproportionately burdensome for the patient. The clinical judgments of the futility of a given therapeutic intervention involve a rational balancing of three factors: efficacy of the given intervention, the purpose of which doctor alone can make; second, the advantage of that intervention, the patients and/or their surrogates can make; and third, the burdens of the intervention (cost, discomfort, pain, or inconvenience), jointly assessed by both physicians and patients and/or their surrogates. Adjusting the relationship among those three criteria is at the heart of prudent, precautionary, and proportionate action [27].

Lastly, the council members on bioethics had opined that the current neurological standard for declaring death, grounded in a careful diagnosis of total brain failure, is biologically and philosophically defensible. The council also concluded that, *in an issue of organ transplantation, determining death and procuring organs should be addressed separately. The questions about the vital status of neurologically injured individuals should be taken up prior to and apart from ethical issues in organ procurement from deceased donors*. The recommendations are: first, to reaffirm the ethical propriety of the dead donor rule (DDR); second, to reaffirm the ethical acceptability of the neurological standard as well as the cardiopulmonary standard; and third, to reject the use of patients in permanent vegetative states as organ donors. The council has concluded that the neurological standard remains valid that was adopted at the President's commission of 1981 and enacted in UDDA.

5.7 Preserving the dead donor rule

The DDR has been secured for the ethical and social acceptability of organ transplantation protocols from their primitive days. This rule demands assurance of the death of the donor as the first step in any ethically legitimate transplantation protocol (other than those involving healthy, living donors). Additionally, the death of the patient must not be accelerated, nor end-of-life care made vulnerable in any way, to accommodate the transplantation protocols [27]. No protocol can demand ethical approval without trustiness to the present rule, in any ethically legitimate transplantation protocol (other than those involving healthy, living donors).

Relaxation of the DDR is a morally and ethically inappropriate and rationally specious way to deal with the uncertainties of the standard for the death of the donor. It leaves the options of the criteria for death to individual preference, amounting to the eventual abolition of any stable criteria for death. Numerous additional dangers are the use of assisted suicide to facilitate organ donation, legitimizing the utilization of patients in permanent vegetative states or of less-than-perfect infants as donors [27]. It exposes "undeclared" patients to "presumed" consent to donation [27, 62, 63].

5.8 Montreal forum (2012): international guidelines for the determination of death

Montreal forum was formed to address the global challenge in response to the request from various countries to "WHO and Transplantation Society" to provide guidance for leading practices and health policy in death determination by neurological and/or circulatory criteria. The guidelines would promote safe practices assuring the absence of diagnostic errors in death determination, safeguarding patients and health care professionals, upgrading public and professional confidence in the dead donation process along with strengthening the availability of organs obtained by ethically legitimate donation and procurement practices. The principles adopted by the forum for discussion were the safeguarding the interests of dying patients overrides facilitating deceased donation for transplantation; task restricted to a scientific, medical, and biological basis for death determination; the principle of the "dead donor rule" applied to deceased donation practices; use of available best scientific and medical evidence for decisions; guidelines and recommendations must have utility, applicability, and be workable in a wide range of global health care practice settings. The key issues of the forum considered death as a biological event with a focus on the physiological aspect of the dying process and death determination and respectfully recognized the impact of attending religious, ethical, legal, spiritual, philosophical, and cultural aspects of death [1].

Forum outcome of the review developed some *terminologies* for clarity for debate/discussion on death. These terminologies reduce a lot of debates on death determination and arriving at a sound consensus; the second outcome was the consensus on *death and brain*, and *death and circulation* regarding illustrative examples of precondition for neurological testing, clinical and laboratory test for diagnosis of neurological and circulatory function, guidelines for declaration, neurological and circulatory sequences of dying process, integrated circulatory and neurological sequences of the dying process, minimum acceptable criteria for clinical test, apnea test and additional lab test, auto-resuscitation, circulatory arrest and brain function, and CPR and life support. These aspect created the foundation for understanding the scientific basis of death declaration and lastly *operational definition of death* with global agreement around complex practice and future research to enrich the knowledge and overcome the gaps. Finally, the forum came to a consensus on seven key areas: Death must be diagnosed on clinical standard based on direct, measurable observation or examination of the patient; physiological events of halting of circulatory and neurological functions leading to death were developed to prove the critical events occurring in a catastrophic injury or illness; clinical tests that satisfy the minimum clinical standard for the death determination were defined for both the neurological and the circulatory sequences; preconditions and confounding situations may impede or invalidate death diagnosis [1]; certain ancillary and/or complementary lab tests might be beneficial in situations where clinical testing cannot be performed or if confounding or special conditions present. Also, the drawback of using some of these tests is acknowledged and further research is recommended to identify the reliability of those tests; a precise terminology regarding death was reviewed and finalized in order to improve clarity in discussions and debate on death determination; the fundamental to define human death should be on measurable biomedical standards and authenticate movement away from anatomically based terms, brain death, or cardiac death misleading to imply the death of that organ. Priority had been placed on the stoppage of neurological or circulatory function and the predominance of brain function for determination of death [1].

The forum came to a consensus on an operational (practical and concrete) definition of human death based on measurable and observable biomedical standards that "Death occurs when there is permanent loss of capacity for consciousness and irreversible loss of all brainstem functions." This might result from permanent stoppage of circulation and/or after catastrophic brain injury. The "permanent" refers to loss of function that cannot resume automatically and will not be returned through intervention. Death is a single phenomenon founded on stoppage of brain function (loss of capacity for consciousness and brainstem reflexes) with two mechanisms to reach that point: permanent absence of circulation or subsequent to a catastrophic brain injury—two entrances, one exit. It is understood that the overwhelming majority of death determination in the world occurs after the stoppage of circulation and usually occurs external to health care settings. In some regions, the dead donation practices include re-establishing circulation (CPR, extracorporeal organ support) following death for the preservation of organs. Future research will enrich this issue for the clarity that constitutes re-establishing circulation, physiologically meaningful circulation, circulation versus oxygenation, and distinctions between organ targeted, regional, and whole-body circulation [1, 64].

6. Variation in death determination criteria in the Asia Pacific

During the 50 years since the publication of reports on the determination of death by neurologic criteria by Harvard University and the WMA (Sydney declaration) in 1968, brain death/death on neurological criteria (BD/DNC) protocols have been developed in many countries around the world. However, some countries still do not have medical standards for BD/DNC, and there is also international and intranational variability between the protocols that do exist [65–68].

Discrepancies were noted in the studies by Wijdicks, Wahlster et al., and Chua et al. between protocols in this region in the criteria used for diagnosis of BD/DNC. Nonetheless, these studies were all limited reviews, though they addressed a number of examiners, observation time, the time between examinations, concordance/discordance with AAN—brain death/death by neurological criteria practice parameters, target value and methods of apnea testing, and requirement for ancillary testing. They did not explore the more distinct aspects of BD/DNC protocols, such as the technique used to rule out the effect of drugs on the evaluation, minimum temperature and blood pressure for an evaluation to be performed, a technique used to assess each component of the examination and findings of BD/DNC, preparation for rationale to discard apnea testing, accepted ancillary tests, need for communication with a person's family, time of death, and stopping of organ support [68]. The existence of a protocol in a given country is dependent on acceptance of BD/DNC as death, access to resources (neurosciences/critical care experts), the presence of a transplant network, and local laws. Religious beliefs markedly influence the acceptance of BD/DNC as death. Although religious views in these countries are distinct from those in the rest of the world, the diversity of political, economic, legal, social, and religious climates throughout the region mirrors that globally [65–68].

A review by Lewisa et al. in 2020 was published in a clinical neurology journal to find out the similarities and differences in the official protocols for the determination of death in Asia Pacific countries (57 of 197 UN) and concluded that protocols for conducting a BD/DNC determination vary markedly. In their report, only 24 of the 37 countries had brain death protocols (69%), but vary in definition such as whole-brain death and brain stem death; a number of examinations vary from single to double, separated by 6–48 hours; and the prerequisites, clinical examination, apnea testing procedure, and indications for/selection of ancillary tests varied. But agreed on that the damage to be irreversible or be permanent, all function/all activities are to be absent before declaring BD/DNC. Also, it is emphasized to harmonize protocols both within this region and worldwide [65–68].

7. Practical guidance for the determination of brain death

Traditionally, death occurs with the confirmation of irreversible cessation of cardiorespiratory function [3, 53–58]. The use of artificial maintenance of life support and organ transplant leads to introduce a new criterion of death determination of permanently nonfunctioning brain, called irreversible coma equated to brain death. In recent years, however, controversy has arisen about the clinical and ethical validity of the neurological standard. **WHO** in 2014 formulated up-to-date update of minimum determination of death criteria for globally acceptable medical practice while respectful to diversities that achieve international consensus on the clinical criteria of determination of death to maintain public trust and promote ethical practices that respect fundamental rights of people and minimize philosophical, ethical, and biomedical debate in the human death. This guideline of clinical criteria on the determination of death suggested that there may be various ways to determine death but there is only one way of being dead, so the two classic algorithms, the brain death and circulatory death, merge into a single endpoint identified as death and should not imply that brain death and circulatory death are two separate phenomena [3]. This guideline also prepares a workable flowchart (**Figure 1**) of **cardiocirculatory algorithm** and **neurological algorithm** to declare the death. The algorithms identify the tests that are required to be conducted at each stage of the event, but they do not specify the details on how each test should be performed, which may be a free-standing documents that do not demand cross-references to other guidelines and be applied to both adults and children, and finally, a checklist is developed to facilitate the implementation of the different components stated in the algorithms [3]. This guideline provides acceptable clinical criteria of medical practice for the determination of death and earn international consensus on death debate but did not mention the detail of the clinical examination [3]. Harvard report describes the clinical criteria, and AAN guidelines on clinical criteria on neurological standard had already been accepted globally.

AAN clinical criteria on the determination of brain death [53–56] can be considered to consist of four steps: Prerequisites, Neurological assessment (coma, absence of brain stain reflex, apnea), Ancillary test, and Documentation.

I. Prerequisites for clinical criteria of brain death determination.

A. Establish permanent and predicted explanation of coma:

- The explanation of coma is often establish by history, clinical examination, neuroimaging, and laboratory tests.
- Rule out the existence of any CNS-depressant drug effect by history, drug screen, calculation of clearance; or, if available, drug plasma levels below the therapeutic range. Prior use of hypothermia (following CPR) may delay drug metabolism. The legitimate alcohol limit for driving (blood alcohol content 0.08%) is a practical threshold below which an examination to determine brain death could adequately proceed.
- Should be no current administration or existence of neuromuscular blocking agents (train of four twitches with maximal ulnar nerve stimulation).
- Should be no critical electrolyte, acid–base, or endocrine disorder (severe acidosis or laboratory values markedly deviated from the norm).

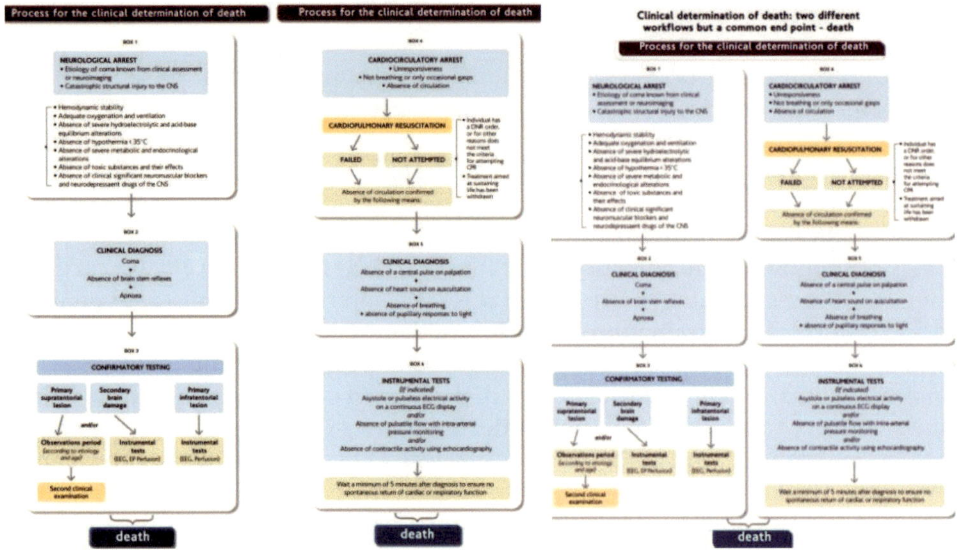

Figure 1.
Source: Ref. [3].

B. Ensure normal core temperature.

Raise the body temperature and to sustain a normal or near-normal temperature (36°C) use a warming blanket. To prevent delaying an increase in $PaCO_2$, normal or near-normal core temperature is preferred during the apnea test.

C. Ensure normal systolic blood pressure. Hypotension or hypovolemia should be corrected by vasopressors or vasopressin. Neurologic examination is commonly reliable with a systolic blood pressure ≥ 100 mm Hg.

D. Perform neurologic examination (one neurological examination is enough to declare brain death in the USA). A certain period of time has to be passed since the onset of the brain insult to rule out the possibility of recovery (usually several hours). However, some US state statutes require two examinations.

E. Legally, all physicians are authorized to determine brain death in the USA. Neurologists, neurosurgeons, and intensive care specialists may have specialized expertise. It appears rational that all physicians making a determination of brain death be absolutely familiar with brain death criteria and have demonstrated competence in this complex examination.

II. Neurological assessment for clinical criteria for brain death determination.

A. Coma:

- Profound loss of consciousness with no response to any stimuli. No evidence of responsiveness. No motor response on noxious stimuli other than spinally mediated reflexes.

B. Absence of brainstem reflexes:

- Lack of pupillary response to bright light is produced in both eyes. Usually, pupils are fixed in a midsize or dilated position (4–9 mm). Constricted pupils signify the possibility of drug intoxication. A magnifying glass can be used in doubtful cases.

- Oculocephalic testing and oculovestibular reflex testing: Absence of ocular movements. Once the integrity of the cervical spine is ensured, the head is briskly rotated horizontally and vertically. No movement of the eyes relative to head movement. The oculovestibular reflex is tested by irrigating each ear with ice water after the patency of the external auditory canal is confirmed. The head is elevated to 30°. Each external auditory canal is irrigated (one ear at a time) with approximately 50 ml of ice water. Eye movement was absent during 1 minute of observation. Both sides are tested, with an interval of several minutes.

- Absence of corneal reflex: Touching the cornea with a piece of tissue paper, a cotton swab, or squirts of water, no eyelid movement will be demonstrated.

- Absence of facial muscle movement to a noxious stimulus: Deep pressure on the supraorbital ridge and the condyles at temporomandibular joints produce no grimacing or facial muscle movement.

- Lack of pharyngeal and tracheal reflexes. This reflex is tested after stimulation of the posterior pharynx with a tongue blade or suction device. The tracheal reflex is most reliably tested by examining the cough response to tracheal suctioning.

C. Apnea test:

Absence of a breathing drive is tested with a CO_2 challenge. Usually, a rise in $PaCO_2$ above normal levels is the typical practice but requires preparation before the test.

Prerequisites for apnea test: (1) Normotension, (2) Normothermia, (3) Euvolemia, (4) Eucapnia ($PaCO_2$ 35–45 mm Hg), (5) Lack of hypoxia, and (6) No prior evidence of CO_2 retention (COPD, excessive obesity).

Procedure:

- Ensure a systolic blood pressure $\geq$ 100 mm Hg, if needed by vasopressors.

- It is mandatory to pre-oxygenate with 100% oxygen for at least 10 minutes to a PaO2 > 200 mm Hg.

- Diminish frequency of ventilation to 10 breaths per minute to eucapnia.

- Diminish positive end-expiratory pressure (PEEP) to 5 cm H_2O (oxygen desaturation with decreasing PEEP suggest problems with apnea testing).

- If pulse oximetry oxygen saturation persists >95%, obtain a baseline blood gas.
- Detach the patient from the ventilator.
- Maintain oxygenation (deliver 100% O_2 at 6 L/min by endotracheal tube).
- Observe closely for 8–10 minutes for respiratory movements. Respiration may be abdominal or may include a brief gasp.
- Exclude if systolic blood pressure decreases to <90 mm Hg.
- Exclude if oxygen saturation measured by pulse oximetry is <85% for 30 seconds.
- If the respiratory drive is absent, repeat blood gas (PaO_2, $PaCO_2$, pH, bicarbonate, base excess) after approx. 8 minutes.
- The apnea test is positive if respiratory movements are absent and arterial PCO2 is ≥60 mm Hg (supports the clinical diagnosis of brain death).
- If the test is inconclusive but the patient is hemodynamically stable during the procedure, the test could be repeated for a longer period of time (10–15 minutes) after the patient is again adequately pre-oxygenated.

III. Supportive tests to diagnose brain death.

The ancillary tests such as EEG, cerebral angiography, nuclear scan, TCD, CTA, and MRI/MRA are at present used for adults in clinical practice. Three tests may be preferred such as EEG, nuclear scan, or cerebral angiogram, as the most hospital has the logistic to perform and interpret. The supportive tests can be done when there is no scope for apnea test or uncertainty exists. The ancillary tests are usually practiced to shorten the duration of the observation period. The interpretation of each of these tests requires expertise. In adults, ancillary tests are not needed for the clinical diagnosis of brain death and cannot replace a neurologic examination.

IV. Documentation of the time of death.

The moment of brain death must be documented in medical records and is the time the arterial PCO_2 reached the target value. But in patients where the apnea test is discarded, the time of death is when the ancillary test has been officially interpreted. A checklist is filled out, signed, and dated.

8. Conclusions

For the millennia, the human has fought with the concept and criteria of and the line between life and death continues to be debated. The profound changes caused by life-sustaining technology and transplantation continue to challenge our notions of life and death. The cardiopulmonary approach is an age-old practice for the determination of death that ensures social acceptance without any debate. The public is also used to rely on the somatic standard for criteria of death such as cooling of the body, absence of breath, loss of consciousness, rigor mortis, putrefaction, and so on.

Despite scientific progress in the last few decades, there remain big variations in the diagnosis criteria applied in each country with legal regulations resulting in misunderstandings among the public and health care professionals. However, the Harvard committee in 1968 develops a set of criteria of the permanently nonfunctioning brain, called irreversible coma equated to brain death. On the same date, the WMA declared a guideline for the determination of death known as the *Sydney declaration*. These two landmarks' innovations change our notions for researching a new challenge in death. The addition of neurological criteria of death to cardiorespiratory criteria of death was a paradigm shift that evolved when patients with acute brain injury could be resuscitated in medical facilities. Brain death, defined as the irreversible cessation of all brain activities, has been included in the medical and legal definition of death for nearly 60 years.

The global philosophical, ethical, legal, and biomedical controversies of determining death due to life support, organ supports, and organ transplantation issues console us in the historic report published (1981) by President commission for the study of ethical problems in medicine and behavioral research, *defining death* that a person is dead, who has sustained either *irreversible stoppage of circulatory and respiratory functions*, or *irreversible cessation of all functions of the entire brain, including the brain stem*. The death determination must be made in accordance with the accepted medical standards [3]. Since then, the neurological standard has been accepted as one of two valid standards for determining death and has been adopted legally in many countries throughout the world. The other accepted standard is the older, traditional cardiopulmonary standard. One of these two valid standards was followed by UDDA legislation for legal criteria of determination of death. President council of bioethics (2008) also reconfirm this definition of death and controlled the DDR rule for organ transplantation purposes. They accept controlled DCD for organ transplantation purposes.

WHO in 2014 published clinical criteria on the determination of death, mentioning various ways to determine death but there is only one way of being dead, so the two classic algorithms of brain death and circulatory death merge into a single endpoint identified as death and should not imply that brain death and circulatory death are two distinct phenomena [3]. They prepare a workable flowchart of cardiocirculatory algorithm and neurological algorithm to declare death. That guideline provides a minimum determination of death criteria to be acceptable for medical practice worldwide to achieve international consensus on clinical criteria to maintain public trust and promote ethical practices that respect fundamental rights of people and minimize philosophical, ethical, and biomedical debate in the human death. The WHO clinical criteria of 2014 did not mention the detail of clinical examination. Harvard report describes the clinical criteria and AAN guidelines on clinical criteria already accepted globally.

American Association of Neurology (AAN) in 2019 validated that brain death is the irreversible loss of all functions of the entire brain and is also equivalent to circulatory death. The testing methods of brain death take into account the age-related anatomical and physiological differences between neonates, infants, and children. Parents and other family members of children undergoing brain death testing may require close attention and additional support [58, 59].

9. Further reading

1. International Guidelines for the Determination of Death—Phase I. Montreal Forum Report; 2012. Available from: http://www.who.int/patientsafety/montreal-forum-report.pdf

2. Clinical Criteria for the Determination of Death. Geneva: World Health Organization; 2017. Licence: CC BY-NC-SA 3.0 IGO

3. Controversies in the Determination of death: A white paper by the President Council on Bioethics (2008). Available from: http://hdl.handle.net/10822/559343

4. Evidence-based guideline update: Determining brain death in adults (Report of the Quality Standards Subcommittee of the American Academy of Neurology). Neurology 2010;74(23):1911–1918. DOI:10.1212/WNL.0b013e3181e242a8

Acknowledgements

All praise to Almighty to bless me with good health to complete this chapter. Acknowledge to all staff of the Forensic Medicine department of IMC and my family member for their sincere cooperation. Special thanks to Prof. Sharfuddin Ahmed, VC of BSMMU for encouraging me to complete. Also thanks to Dr. Ferdausi Rahman for her review of some portion of scripts.

Authors' contributions

Prof. Md Shah Alam developed the conception and design of the article and drafted the manuscript, providing important intellectual content.

Conflict of interest

The authors declare no conflict of interest.

Ethical approval

Not applicable.

Abbreviations

AAN	American Association of Neurology
BD	brain death
CNS	central nervous system
CTA	computed tomography angiography
CDD	controlled death donor
COPD	chronic obstructive pulmonary disease
DDR	death donor rule
MRA	magnetic resonance angiogram

PVS	persistent vegetative state
TCD	transcranial Doppler
CCA	cerebral circulatory arrest
BSMMU	Bangabandhu Sheikh Mujib Medical University, Bangladesh
DNC	death on neurological criteria
UDDA	unifying determination of death act
WHO	World Health Organization
WMA	World Medical Association/World Medical Assembly

Author details

Md. Shah Alam Panna[1,2,3]

1 Department of Forensic Medicine, Bangabandhu Sheikh Mujib Medical University (BSMMU), Dhaka, Bangladesh

2 Department of Forensic Medicine and Toxicology, Ibrahim Medical College, Dhaka, Bangladesh

3 Department of Forensic Medicine and Toxicology, Sher-E-Bangla Medical College, Barishal, Bangladesh

*Address all correspondence to: drpanna@yahoo.com

References

[1] International Guidelines for the Determination of Death—Phase I. Montreal Forum Report; 2012. Available from: http://www.who.int/patientsafety/montreal-forum-report.pdf [Accessed: 17 May 2021]

[2] Naffine N, Richards B, Rogers W. Scaring us all to death: The need for responsible legal scholarship on post-mortem organ donation. Journal of Law and Medicine. 2009;**16**:696-707

[3] WHO (2017). Clinical Criteria for the Determination of Death. WHO technical expert consultation. Geneva, Switzerland, WHO Headquarters: World Health Organization; 22-23 Sept 2014. https://apps.who.int/iris/handle/10665/254737 [Licence: CC BY-NC-SA 3.0 IGO]

[4] Shemie SD. Clarifying the paradigm for the ethics of donation and transplantation: Was dead really so clear before organ donation? Philosophy, Ethics, and Humanities in Medicine. 2007;**2**:18

[5] Gardiner D, Shemie S, Manara A, Opdam H. British Journal of Anaesthesia. 2012;**108**(S1):i14-i28. DOI: 10.1093/bja/aer397

[6] Jakušovaitė I, Luneckaitė Ž, Peičius E, Bagdonaitė Ž, Riklikienė O, Stankevičius E. Review—Determination of death: Metaphysical and biomedical discourse. Medicina. 2016;**52**(4):205-210. DOI: 10.1016/j.medici.2016.06.002

[7] Sarbey B. Definitions of death: Brain death and what matters in a person. Journal of Law and the Biosciences. 2016;**3**(3):743-752. DOI: 10.1093/jlb/lsw054

[8] David D. Definition of death. In: Zalta EN, editor. The Stanford Encyclopedia of Philosophy. 2017. Available from: https://plato.stanford.edu/archives/spr2017/entries/death-definition/ [Accessed: 16 May 2021]

[9] Lock M. Twice Dead: Organ Transplants and the Reinvention of Death. California, USA: University of California Press; 2002. Available from: http://www.jstor.org/stable/10.1525/j.ctt1pn7ph [Accessed: 27 May 2021]

[10] Locke J. In: Winkler KP, editor. An Essay Concerning Human Understanding (Abridged Edition). Indianapolis: Hackett Publishing; 1996. p. 9

[11] Lizza JP. The conceptual basis for brain death revisited. In: Machado C, Shewmon DA, editors. Brain Death and Disorders of Consciousness (Advances in Experimental Medicine and Biology). Vol. 550. Boston, MA: Springer; 2004

[12] Condic ML. Determination of death: A scientific perspective on biological integration. The Journal of Medicine and Philosophy. 2016;**41**(3):257-278. DOI: 10.1093/jmp/jhw004

[13] Kushner T. Having a life versus being alive. Journal of Medical Ethics. 1984;**10**:5-8

[14] Schofield GM, Urch CE, Stebbing J, Giamas G. When does a human being die? QJM. 2015;**108**:605-609

[15] Emmerich N. Commentary—The Deaths of Human Beings. 2015. Available from: https://academic.oup.com/qjmed/article/109/4/229/1752767 [Accessed: 03 May 2021]

[16] Ayodeji O. Christians' perception of the concepts of death and judgment: A multimodal discourse analytical study of selected editions of Christian Women Mirror Magazine. International Journal of English and Literature. 2013;**4**(10):508-515. DOI: 10.5897/IJEL2013.0446

[17] Plato G. Republic (524b-c) [trans. Robin Waterfield]. New York: Oxford University Press; 2008. p. 131

[18] Korein J. Brain death. In: Cotrell JE, Tundorf H, editors. Anaesthesia and Neurosurgery. St. Louis: C. V. Mosby; 1980. pp. 282-432

[19] Korein J. The problem of brain death: Development and history. Annals of the New York Academy of Sciences. 1978;**315**:19-38

[20] Korein J. Ontogenesis of the fetal nervous system: The onset of brain life. Transplantation Proceedings. 1990;**22**: 982-983

[21] Machado C, editor. Brain Death: A Reappraisal. New York: Springer Science; 2007

[22] Machado C, Sherman DL. Brain Death and Disorders of Consciousness. Vol. 50. New York: Kluwer Academics/ Plenum; 2004

[23] Machado C. Death on neurological grounds. Journal of Neurosurgical Sciences. 1994;**38**:209-222

[24] Machado C. Consciousness as a definition of death: Its appeal and complexity. Clinical Electroencephalography. 1999;**30**:156-164

[25] Bernat JL. The concept and practice of brain death. Progress in Brain Research. 2005;**150**:369-379. DOI: 10.1016/S0079-6123(05)50026-8

[26] Beecher HK. Ethics and clinical research. The New England Journal of Medicine. 1966;**274**:1354-1360. DOI: 10.1056/NEJM196606162742405

[27] Controversies in the Determination of Death: A White Paper by the President's Council on Bioethics. Washington, DC. 2008. Available from: http://hdl.handle.net/10822/559343

[28] McAlister V. Maimonides's cooling period and organ retrieval. Canadian Journal of Surgery. 2004;**47**(1):8-9

[29] Whetstine LM. The history of the definition(s) of death: Form 18th century to the 20th century. In: Crippen DW, editor. End-of Life Communication in ICU. New York, NY: Springer; 2008. pp. 65-78

[30] Alexander M. The rigid embrace of the narrow house: Premature burial and the signs of death. Hastings Center Report. 1980;**10**:27

[31] Ewin RE. Reasons and the Fear of Death. Maryland: Rowman & Littlefield; 2002. p. 109

[32] Harvey. Exercitatio anatomica demotucordis et sanguinis in animalibus. 1628. Available from: http://www.fordham.edu/halsall/mod/1628harvey-blood.html

[33] Loukas M, Tubbs RS, Shoja MA, Apaydin N, Lam R. Ibn al-Nafis (1210-1288): The first description of pulmonary circulation. American Surgeon. 2008;**74**(5):440-442. DOI: 10.1177/000313480807400517

[34] Ibn an-Nafīs. Muslim Physician. Britannica; 2021. Available from: https://www.britannica.com/biography/Ibn-an-Nafis [Accessed: 6 June 2021]

[35] Britannica. The Editors of Encyclopaedia. "Ibn an-Nafīs". Encyclopedia Britannica. 2021. Available from: https://www.britannica.com/biography/Ibn-an-Nafis [Accessed: 25 May 2021]

[36] Bondeson J. Buried Alive: The Terrifying History of Our Most Primal Fear. London: W.W. Norton; 2002

[37] Guernsey. Real and apparent death. In: Homeopathic Domestic Practice. New York: William Radde; 1853. pp. 493-514. Available from: http://books.

google.co.uk/ebooks/reader?id=ZDXXwnGXA1YC&printsec=front cover&output=reader

[38] Powner DJ, Ackerman BM, Grenvik A. Medical diagnosis of death in adults: Historical contributions to current controversies. Lancet. 1996;**348**: 1219-1223

[39] Australian Museum. Stethoscopes. Available from: http://australianmuseum.net.au/Stethoscopes/

[40] Mollaret P, Goulon M. Coma de´passe´ (preliminary memoir). Revista de Neurologia. 1959;**101**:3-15

[41] Wijdicks EFM. Brain Death. Philadelphia: Lippincott Williams & Wilkins; 2001

[42] Pius XII. The Prolongation of Life [An Address of Pius XII to an International Congress of Anesthesiologists, Nov. 24, 1957]. Vatican City: The Pope speaks; 1958. pp. 393-398

[43] Wertheimer P, Jouvet M, Descotes J. Diagnosis of death of the nervous system in comas with respiratory arrest treated by artificial respiration. Presse Médicale. 1959;**67**:87-88

[44] Fischgold H, Mathis P. Obnubilations, comas et stupeurs. Electroencephalography and Clinical Neurophysiology. 1959;(Suppl 11):1-124

[45] Wijdicks EF. The neurologist and Harvard criteria for brain death. Neurology. 2003;**61**:970-976

[46] Tham SJ. Harvard brain death criteria and organ transplantation: A historical revisit. Studia Bioethica. 2009;**2**(2):40-50

[47] Beecher HK, Adams RD, Barger AC, Currn JW, et al. A definition of irreversible coma. Report of the Ad Hoc committee of the Harvard Medical School to examine the definition of brain death. Journal of the American Medical Association 1968;**205**:337-340. DOI:10.1001/jama.1968.03140320031009

[48] Gilder SSB. Twenty-second World Medical Assembly. British Medical Journal. 1968;**3**:493-494

[49] Machado C, Korein J, Ferrer Y, Portela L, de la C Garcı'a M, Chinchilla M, et al. The declaration of Sydney on human death. Journal of Medical Ethics. 2007;**33**:699-703. DOI: 10.1136/jme.2007.020685

[50] De Georgia MA. History of brain death as death: 1968 to the president. Journal of Critical Care. 2014;**29**(4):673-678. DOI: 10.1016/j.jcrc.2014.04.015

[51] Presidents commission for the study of ethical problems in medicine and biomedical and behavioral research defining death: Medical, legal and ethical issues in the determination of death [herein after commission]. 1981. pp. 13-16. Available from: https://academic.oup.com/jlb/article/3/3/743/2327864 [Accessed: 13 November 2020]

[52] Plum F, Posner JB. The Diagnosis of Stupor and Coma. 3rd. ed. Philadelphia: F. A. David Co.; 1980. pp. 6-7

[53] Wijdicks EFM, Varelas PN, Gronseth GS, Greer DM. Evidence-based guideline update: Determining brain death in adults (Report of the Quality Standards Subcommittee of the American Academy of Neurology). Neurology. 2010;**74**(23):1911-1918. DOI: 10.1212/WNL.0b013e3181e242a8

[54] Nakagawa TA et al. Guidelines for the determination of brain death in infants and children: An update of the 1987 task force recommendations—Executive summary. Annals of Neurology. 2012;**71**(4):573-585

[55] Pope T. Brain death and the law: Hard cases and legal challenges. The

Hastings Center Report. 2018;**48** (Suppl 4):S46-S48

[56] Russell JA, Epstein LG, Greer DM, Kirschen M, Rubin MA, Lewis A. Brain death, the determination of brain death, and member guidance for brain death accommodation requests: AAN position statement. Neurology. Jan 2019;**92**(5)228-232. DOI: 10.1212/wnl.0000000000006750

[57] Nakagawa TA, Ashwal S, Mathur M, Mysore M, Committee for Determination of Brain Death in Infants Children. Guidelines for the determination of brain death in infants and children: An update of the 1987 task force recommendation—Executive summary. Annals of Neurology. 2012;**71**(4):573-585. DOI: 10.1002/ann.23552

[58] Rukovets O. Pediatric Brain Death Determination Protocols Vary from State to State. AAN Publication; 2021. Available from: http://Journals.Iww.com [Accessed: 8 July 2021]

[59] Aimee M, Melissa U. Brain death: History, updates, and implications for nurses. American Journal of Nursing. 2020;**120**(3):32-38. DOI: 10.1097/01.NAJ.0000656332.62081.98

[60] Zamperetti N, Bellomo R. Total brain failure: A new contribution by the president council on bioethics to the definition of death according to neurological standard. Intensive Care Medicine. 2009;**35**:1305-1307. DOI: 10.1007/s00134-009-1499-z

[61] Zamperetti N, Bellomo R, Defanti CA, Latronico N. 1968-2003: Irreversible apnoeic coma 35 years later. Towards a more rigorous definition of "brain death". Intensive Care Medicine. 2004;**30**:1715-1722

[62] Pellegrino ED. Decisions at the end of life: The use and abuse of the concept of futility. In: De Dios Vial Correa J, Sgreccia E, editors. The Dignity of the Dying Person (Proceedings of the Fifth Assembly of the Pontifical Academy for Life). Citta del Vaticano: Libreria Editrice Vaticano; 2000. pp. 219-241

[63] Brugger EC. D. Alan Shewmon and the PCBE's white paper on brain death: Are brain-dead patients dead? The Journal of Medicine and Philosophy. 2013;**38**:205-218. DOI: 10.1093/jmp/jht009

[64] Shemie SD, Homby L, Baker A, et al. International guideline development for the determination of death. Intensive Care Medicine. 2014;**40**:788-797. DOI: 10.1007/s00134-014-3242-7

[65] Wahlster S, Wijdicks EF, Patel PV, Greer DM, Hemphill JC 3rd, Carone M, et al. Brain death declaration: Practices and perceptions worldwide. Neurology. 2015;**84**:1870-1879

[66] Wijdicks EF. Brain death worldwide: Accepted fact but no global consensus in diagnostic criteria. Neurology. 2002;**58**:20-25

[67] Chua HC, Kwek TK, Morihara H, Gao D. Brain death: The Asian perspective. Seminars in Neurology. 2015;**35**:152-161

[68] Lewisa A, Liebmanb J, Bakkarb A, Kreiger-Bensonb E, Kumpfbeckb A, Sam D, et al. Determination of brain death/death by neurologic criteria in countries in Asia and the Pacific. Journal of Clinical Neurology. 2020;**16**(3):480-490. DOI: 10.3988/jcn.2020.16.3.480

Chapter 3

Compassion Versus Care in Healthcare Institutions: What's the Difference?

Una P. Canning

Abstract

In February 2013, the Francis Report outlined what it described as 'systematic failings' at Mid Staffordshire NHS Foundation Trust resulting in the death and suffering of many patients through neglect (in the UK context, hospitals can apply to gain foundation trust status. Foundation trust hospitals are part of the National Health Service (NHS) but are not directed by central government and have greater freedom to decide the way services are delivered. They adhere to core NHS principles of free medical treatment based on need and not the ability to pay.) A lack of compassion, particularly among nursing staff, was identified as one of the contributing factors to poor care. The NHS was founded on the core value of compassion that today is one of six values all NHS staff are expected to demonstrate. Frequently invoked as a means to ensuring good patient care, it is a concept that is contested by a number of writers who argue that such moral emotions are not only unnecessary but dangerous. The purpose of this work is to explore the difference between compassion and care (but not medical treatment) in the context of the NHS. The paper draws on the work of Anca Gheaus, who argues there is a distinction to be made between the two and that while it is possible to be compassionate towards everybody, the ability to care, is limited to fewer people and is a more intense and engaged activity. Regarded as the founding myth of the NHS, the work also draws on the parable of the Good Samaritan to make the distinction between the two concepts more visible, and argues the roles played by the Good Samaritan and the innkeeper, remain relevant to the workings of today's healthcare system. It also reflects on the need for kindness within the system.

Keywords: Care, compassion, Francis Report, Good Samaritan, NHS

1. Introduction

In February 2013, the Francis Report [1] outlined what it described as 'systematic failings' at Mid Staffordshire NHS Foundation Trust (foundation trusts are still part of the National Health Service (NHS) but are not directed by central government and have greater freedom to decide the way services are delivered) [2]. These systematic failings resulted in the death and suffering of many patients through neglect. In looking to identify the causes of these failings, a lack of compassion particularly among nursing staff, was identified as one of the contributing factors. In his response to the Francis Report, the prime minister, David Cameron, recommended nurses 'be hired and promoted on the basis of having compassion as a vocation' and not just academic qualification to 'ensure that it plays a part in every healthcare interaction [3]. Cameron's suggestion that nurses' pay should be made dependent on their ability to demonstrate compassion in their jobs, gives rise to what Anne Bradshaw [4] describes as 'a McDonald's – type nursing care rather than heartfelt care.' In Bradshaw's view [4], failures in the NHS cannot be attributed to failures in nursing care but is arguably the result of the rejection by a pluralistic society of 'Judaeo-Christian values' that no longer regards such values as relevant or valuable. In their absence, a utilitarian model of healthcare has emerged that is 'market-driven and bureaucratised' and is an approach that has overtaken the value of care [4]. Following an economic downturn in the 1970s ideas from economics started to have an impact on medicine. Economists such as Alan Maynard advanced a utilitarian argument that criticised healthcare delivery in the UK as an inefficient use of resources, claiming it was unethical because it deprived other patients of services [5].

According to Bradshaw [4] 'Judaeo-Christian virtues, such as compassion, [...] used to play a significant, though often unstated part in medicine and nursing in the western world' but this moral framework came to be rejected by nursing leaders who 'were anxious to remove the quasi-religious base for care.' The change first occurred in the USA but later began to influence UK nursing that up until the 1970s saw generations of student nurses trained using the classic nursing textbook written by Evelyn Pearce [4]. In her writings, Pearce emphasised the importance of nurses developing 'a moral character' and the need to exercise 'kindness, compassion and unselfishness which contact with sick people demands.' [3] As a result of these changes, UK nurses entering training with a presumption of a moral framework akin to that emphasised by Pearce 'underwent a professional socialisation and doctrinal conversion that repudiated such values.' [4] Instead, nursing began to be influenced by contemporary views of care emerging from the USA that included the works of Carole Gilligan and Nell Noddings. Both these writers espoused a view of feminine morality that countered the Kantian view of rights and justice, and resisted the ethics of care being formalised into abstract principles [6]. In the case of Noddings, she regarded the moral impetus to care as:

> *'An engrossed subjective experience and not a moral norm. It is neither generalizable nor universalizable and depends on an affirmative response in the cared-for. From this perspective, "compassion" is an emotional response dependent on reciprocation, and not a virtue to be cultivated as an aspect of individual character.' [4].*

For Bradshaw [4] writings on the 'feminine ethic of care' is problematic as it provides no 'moral basis for the nurse to help the unresponsive, indifferent or even hostile and unsympathetic stranger. Not normative, and derived from the feminine nature, it is problematic for the male nurse too.' Anca Gheaus [6] agrees and argues the ethics of care is not a women's morality but a universal morality and 'the ability to care well are things in which women and men can (and should) be socialised' and if conceived as an exclusively feminine morality, will only lead to 'exclusion, oppression and neglect.' Other criticisms by feminist writers against feminist moral reasoning in terms of care have also been raised. A typical worry about such an ethics of care is noted by Claudia Card:

> *'Resting all of ethics on caring threatens to exclude as ethically insignificant our relationships with most people in the world because we do not know them, and we never will. Regarding as ethically insignificant our relationships with people remote from us is a major constituent of racism and xenophobia.' [7].*

In this work I intend to explore these two views in the context of healthcare. In doing so, I hope to signpost a way though these opposing views; that is, between the view that moral emotions such as compassion are necessary for delivering good healthcare and those who oppose such views and argue compassion is irrelevant in healthcare settings because 'it is an engrossed, subjective experience.' To do so, I explore the concepts of compassion and care and rely on the work of Anca Gheaus [6] who argues that there is a distinction to be made between other regarding virtues such as compassion and that of care. In the work I also refer to the parable of the Good Samaritan and argue that the respective roles played by two of the key characters, the Samaritan and the innkeeper, are a good example of this distinction.

Today, compassion is included as one of six values in the NHS Constitution [8] and is frequently invoked as a means for ensuring good patient care. Despite its inclusion, it is a concept that is contested by a number of writers who argue such moral emotions are not only unnecessary but dangerous [3]. In the literature, care has defined in general terms as a disposition and activity of meeting needs [9] but according to Gheaus, although care is similar to other moral emotions such as compassion, it is a more intense and engaged activity [6]. Recounting the parable of the Good Samaritan - considered to be the prototype of the British welfare state - the paper reflects upon the respective roles played by the Samaritan and the innkeeper in the parable as a means to illuminating the difference between these two concepts. The paper proposes that the concept of care might be a more useful means of safeguarding good patient care because it involves the desire to actively help when possible whereas compassion is more of an attitude one may have towards people in general. The argument developed here references the findings of the Francis Report [1] and the failures in care identified at Mid Staffordshire NHS Foundation Trust. Obstacles to providing good care are discussed in the final sections of the paper.

2. My brother's keeper

In the 1970s the British Secretary of State for Health, Barbara Castle, stated that the NHS 'is the nearest thing to the embodiment of the Good Samaritan that we have in any aspect of our public policy.' [10] In 1970 Richard Titmuss published his study of British blood donors, *The Gift Relationship* and in it he describes how the universal impulse to help strangers was simply enacting a fundamental truth of human existence and 'to love oneself, one must love strangers.' [11] In Bradshaw's view [4] this specific understanding of care in nursing differs from contemporary understandings and in 'a modern supposedly secular and plural society' it is disingenuous to claim its values and says there is a need to acknowledge that new models of care have different underpinning values. The NHS Constitution currently emphasises six values that staff are expected to demonstrate as part of their work including: 'Working together for patients; compassion; respect and dignity; improving lives; commitment to quality of care; everyone counts.' [8] According to the Constitution, compassion is central to the care provided in the NHS and is achieved by:

> *'Respond[ing] with humanity and kindness to each person's pain, distress, anxiety or need. We search for the things we can do, however small, to give comfort and relieve suffering. We find time for patients, their families and carers, as well as those we work alongside. We do not wait to be asked, because we care.' [8].*

Historically, the development of the compassionate character provided the nursing profession with its ethos until its rejection in the 1970s. Up until then, nurse training involved becoming kind and compassionate, as well as becoming technically competent with 'the character of the nurse considered just as important as the knowledge she possesses.' [4] But with the traditional system of nurse training considered no longer acceptable, nursing leaders set the profession on 'an entirely new course.' [12] In 1986 the UK Central Council for Nursing, Midwifery and Health Visiting (UKCC) published proposals that saw the implementation of *Project 2000* and the traditional apprenticeship style of nurse training giving way to a model that saw nurses acquire student status. With these changes, concerns were raised that nursing had become too academic and had 'ditched its core vocation to care.' Nurses it was claimed, had become 'too posh to wash' and the bedpan - 'the enduring symbol' of traditional nursing was now being emptied by someone else [12].

In Bradshaw's view [4] the term 'care in nursing' has traditionally been understood as an axiom and is a normative moral practice of compassionate help for the stranger in need. Bradshaw believes this is the understanding the UK government, the Royal College of Nursing and the NHS Confederation appear to presuppose in their discussion of the subject and is a view underpinned by the parable of the Good Samaritan that has its origins in the Old Testament. In the ancient *Book of Leviticus*, Israelites are commanded to welcome strangers and told to: 'treat resident aliens as though they were native born and love them as yourself.' [13] In the New Testament the parable of the Good Samaritan also recalls this command. It recounts the story of a traveller attacked by bandits and left to die on the roadside and the response of the four characters who encounter the victim including the Samaritan (despised by the Jewish people), the priest (a privileged member of Jewish society), the Levite (a lawyer, inferior to the priest but still belonging to a privileged group in Jewish society) and the innkeeper (despised by Jews and Samaritans alike.) [10] Drawing on religious and racial tensions common among Jews and Samaritans at that time, the story tells of how two esteemed pillars of Jewish society (the priest and the Levite) refuse to help the victim as he lay gravely injured by the roadside because of fear of attack and also fear of being defiled through touching the victim. Instead, it is the Samaritan who comes to the victim's aid and as he dresses the victim's wounds, flouts laws about ritual impurity common among both Jews and Samaritans. During the course of the story, listeners are also told that the Samaritan was a seller of oil and wine, an occupation both Jews and Samaritans despised, as such traders were considered 'very shady people, indeed, even criminals.' [10] Because of this listeners would have been surprised at the Samaritan's actions and of his 'willingness to go to the margins of society in his ministry of healing [that] defined the depth of his compassion.' [10] The role played by the fourth character, the innkeeper, (a class of people generally considered to be thieves and robbers) adds a further unexpected twist to the story as Arbuckle explains:

> *'The inn in the story was a den of thieves, and the head thief was the innkeeper, yet he was prepared to help the victim, for a price [...] Knowing what to expect from the innkeeper, the Samaritan simply bribed him in order to guarantee that the patient would be looked after and kept alive. He left the innkeeper a certain amount but promised more when he returned.' [10].*

According to Arbuckle the Samaritan was a 'shrewd businessman' and as well as having the material goods to help the victim, he also had the relevant management skills as was evident in his dealings with the innkeeper. Alert to the weaknesses of human nature, the Samaritan bribed the innkeeper so he would care for the victim and is an example says Arbuckle, of the values of efficiency and excellence. For Arbuckle, there is a convergence between the values inherent in the Good Samaritan parable and the founding values of the welfare state in Britain and he identifies two kinds of values (final and instrumental) that are relevant to both:

> *'Final, that is, meaning a desired end state, and [...] instrumental, that is, those actions that are adequate or essential to achieve the desired end.' [10].*

3. Reasoning from love to the moral inclusion of strangers

In examining the relationship between compassion and care Bradshaw [4] suggests the word compassion is associated with religious belief, especially Christianity and that caring as an activity requires 'cultivating the virtue of compassion in the carer' that in the context of nursing, is acquired through training. On this view, compassion is understood as suffering together with another and is more than an emotion or feeling but a whole praxis that has 'a moral and intellectual component that is universalisable.' [4] For Bradshaw, compassion is not an abstract theoretical idea but is lived out in the practice of the carer - a view also held by Florence Nightingale.

But in the context of healthcare, Anna Smajdor [3] argues that such moral emotions are not only unnecessary but also dangerous because our capacity to love and feel compassion is so circumscribed that '[U]nless we regard healthcare professionals as saints, we cannot demand that they guarantee an unlimited flow of compassion for each patient' and that:

> *'Medical professionals need to protect themselves as well as performing their medical duties, and if we demand compassion in addition to medical expertise and knowledge, we are setting our healthcare professionals up for failure.' [3].*

In Smajdor's view the terminology surrounding the word care is unhelpful because '[T]o care can be either to feel a certain way, or to carry out certain activities' and compassionate care is only really possible when, for example, we are a mother caring for her child or a wife for her husband. This kind of 'relational' care ethic is all very well when we are treating loved ones but as healthcare professionals do not routinely treat loved ones, it is dangerous and unfair relying on compassion as the motivation for ensuring essential tasks are carried out.

Anca Gheaus in her work, rejects the view that the relational aspect of care is a barrier to the moral inclusion of strangers and argues the relational nature of human beings has an epistemic role to play in defining the scope of human morality [6, 7]. Acknowledging that allowing personal relationships and emotions to be part of the argument that informs morality runs the risk of treating those unrelated to us unfairly, Gheaus argues it can still be a source of value in determining the proper scope of our morality as it is:

> *'Imaginable, that [...] intelligible, emotional connections (which I call here 'love') based on the universal need people have for each other, can do the work left unfinished by the argument from actual connectedness.' [7].*

Defining love as 'personal' and directed towards a particular individual Gheaus argues this type of love can form the basis of our ability to respond morally to strangers. It is an argument she claims that is not dissimilar to religious love for humanity (which is universal and impartial) as it too, has sometimes 'been considered to facilitate an ability to see the equal worth of all human beings.' [7] By invoking people's relationships and need for each other, Gheaus argues we can engage in moral reasoning because of the relational fact that we are creatures who need to love others and also need others' love.' And although we relate differently to loved ones compared to strangers, she argues we should be morally concerned for strangers because they are 'at least potentially – *somebody's* loved ones.' [7].

In reasoning from love to morality Gheaus [7] draws on feminist ethics, including the works of Sara Ruddick (1989) [14] and Eva Kittay (1999) [15] both of whom 'indicate a way of reaching universalising moral conclusions from the existence of particular, personal bonds of love.' She also draws on the work of the philosopher

Raimond Gaita [16] and his book *A Common Humanity* (2000) where he invites us to think of how our commitments to those we love, are relevant for the obligations we have towards other people in general. In his work, Gaita argues that precisely because we are able to love some individual human beings, we are able to gain a full understanding of the moral value of people in general and he illustrates his point with the story of a nun's visit to the wards of a psychiatric hospital where he worked for a period in the 1960s. In the story, Gaita describes the moral responses of the doctors and nurses looking after the patients as diverse, ranging from brutal to kind. He also tells of how despite the expression of compassion by some of the regular staff, it was only the visiting nun that related to the psychiatric patients as equals and in doing so, acknowledged their full humanity. This came as a revelation to Gaita, who believed the nun's attitude was made possible by her love for the patients – in this case universal, Christian or saintly love. But once revealed Gaita believed this love was independent of the nun's religious background and accessible also to those who do not hold metaphysical beliefs.

Agreeing with Gaita that love has an epistemic role to play in our morality, Gheaus however considers the account of the nun's revelatory love as problematic as a means to universal moral inclusion because it is potentially unsustainable. Instead, Gheaus argues that an example of secular love, that is personal love directed at a particular individual, would be more convincing because the nun's love can best be understood as impartial and unconditional and therefore more sustainable as a general attitude than it would be coming from any of the nurses or doctors involved in caring for the patients. For Gheaus, it is not mere coincidence that the daily hands on care for the patients was done by the non-compassionate nurses, and the ongoing responsibility borne by the compassionate but condescending doctors while the nun was just a passer-by. Gheaus' argument here is that in the context of ordinary, everyday challenges, striving to maintain unconditional love is difficult, if at all humanly possible, and impartial love, disentangled from knowledge of the particularities of the beloved, is more easily amenable to being unconditional. On this view, an example of ordinary, partial and therefore more fragile love would be more convincing for two reasons:

> *'First because this is the love which most of us experience. And second, because as already noted, this ordinary love unlike impartial and unconditional Christian love is not as such a moral emotion.' [7].*

A further feature identified by Gheaus of personal, partial love that is constitutive of our human morality, is that of beings 'who need each other', and whose 'moral agency is in part determined by our need to be in (loving) relationships.

In her discussion of human need, Onora O'Neill argues that utilitarian thinking assigns no special importance to human need and leaves vital dilemmas unclarified and unresolved, despite the fact that all human action is predicated on 'a plurality of mutually vulnerable beings who never achieve more than limited and specific forms of rationality, independence and self-sufficiency.' [17] Developing an abstract Kantian argument that rejects 'principled indifference to others' she argues for a theory of obligations similar to that found in Christian and other religious traditions and also 'present in the idiom of much of our social life.' [18] In contrast to a utilitarian perspective that endorses the pursuit of happiness without specific concern to meet needs, or a human rights perspective that often fails to allocate obligations to help those in need, O'Neill proposes an obligations theory that is premised on not 'bas[ing] our lives on principles that are indifferent to, or neglectful of others.' [18] On this view 'the fact that we cannot help everyone only shows that we have no obligation to help everyone and not that we have no obligation to help anyone.' [18]

According to O'Neill 'ethical traditions that extol universal benevolence, love for all mankind, or concern for all' are misleading because nobody can provide help or care for all others and therefore the rejection of indifference cannot be expressed in action for all others [18]. As help or care for all others is not possible, the rejection of indifference is demonstrated through the provision of '*some* care to sustain *some* others in *some* ways' [18] that is not trivial or sporadic and sustains at least some of their capacities and capabilities.

Not contingent on any special relationship, these obligations are called *imperfect obligations* because concern for all is not possible and therefore selective and 'the pattern and occasions of virtuous action may leave much open for judgement.' [18] Regarded as a moral duty, *imperfect obligations* cannot be claimed as a matter of right and are distinct from *perfect obligations* which gives the right to one party to take legal action against a party that has failed to perform a particular duty. In O'Neill's view, contemporary liberal thinking marginalises imperfect obligations and excludes all but justice from their ethical perspective and takes pride in being 'agnostic about the good of man.' [17] If we want to establish intellectually robust norms in health, O'Neill suggests it would be preferable to start from a systematic account of obligations, rather than of rights because it makes it easier to spot incoherence in the system [17].

4. Compassion versus care: what's the difference?

For Gheaus the terms love and care partially overlap. In the literature 'care' has been defined in general terms as a disposition and activity of meeting needs [9]. In her analysis Gheaus [6] argues that when thinking of the word care there are several understandings of the word in the literature including: care as a type of work, care that signifies a special emotional bond between persons, and care as a virtue – a type of moral motivation as in 'caring about.' Distinguishing between several possible concepts of care in the literature and their relationship to each other, Gheaus [6] argues the different concepts do not necessarily exclude each other but that each presupposes the others to some extent. Rather than attempt to reduce the various meanings of the word found in the literature to a single concept, Gheaus [6] argues for a multi-layered understanding of care and proposes that we can best understand care and its moral significance, by connecting it to the idea of needs. The adoption of a multi-layered understanding that is connected to the idea of needs, makes it possible in Gheaus' view, to identify different contexts of care such as healthcare 'that are not based on care as an emotion close to love' but depending on context, can enable us to care for distant others [without] any emotional connection towards those one is embracing.' [6] In the many different senses in which the word care is used in the feminist literature, Gheaus argues the concept of need is a common feature and that:

> *'The most widespread way of understanding care is responsiveness to the needs of concrete individuals. The moral value of care is intimately linked to the fact that human beings are most of the time not self-sufficient, invulnerable creatures, but beings who depend on others for survival and thriving.' [6].*

For Gheaus meeting the needs of others necessitates individuals being treated in a personal way and this requires an interest in, and knowledge of, the particular circumstances of each person.

In the literature discussions about care have mainly been associated with compassion and benevolence. While noting how similar care is to other moral emotions, authors have rarely provided an exact analysis of how they differ and according to Gheaus [6] this has precluded a full understanding of the distinctiveness of care. In attempting to delineate the boundaries between care and other moral emotions, Gheaus [6] argues that the particular meaning one attaches to the word means it can also come close to other moral emotions such as benevolence, compassion, empathy but that the scope of care is wider and involves acting on behalf of others. For Gheaus care is distinct from compassion or pity because one can be compassionate towards everybody but to qualify as care, the desire to actively help must also be present:

> *'Compared with pity, compassion or charity, the scope of care is wider. Both pity and compassion are mainly about concern with people's suffering and desire that it should be alleviated. But [...] alleviating harm is only part of the work of care. Care is as much concerned with fostering growth and happiness; it is as appropriate a reaction to cheerful situations as to distressful ones. An additional point is that one may be called compassionate or feel pity without necessarily getting too actively involved with the suffering. By contrast, to care for someone who suffers (or rejoices) requires a higher degree of commitment than that of compassion. To qualify as "care", an attitude must involve at least the desire to actively help when possible.' [6].*

Distinguishing care from altruistic motives that typically targets strangers, she develops the argument further by taking the example of benevolence:

> *'One can be truly benevolent without being committed to act extensively on behalf of those who are the objects of benevolence. Even more important, benevolence is a much less partial disposition than care. One can perhaps be benevolent towards everybody or at least towards everybody one interacts with [...] but care is more intense and engaged [and] our capacity to give it is limited to fewer people.' [6].*

In Arbuckles [10] analysis of the Good Samaritan parable, the distinction between the two concepts can be readily identified. Motivated by compassion for the victim, the Samaritan assists him at the roadside but is unable to commit to looking after the victim's long-term needs because of business commitments: unable to stay and provide the necessary care, he uses the resources at his disposal (money) to pay someone else to do the caring. That 'someone else' was the innkeeper who agreed to actively help the victim for money and in return, was rewarded for his efforts with the promise of further payment on the Samaritan's return.

In thinking of the relevance of the Good Samaritan parable to contemporary healthcare, the role of abstraction is important for without abstraction 'there is no communication with those of differing cultures [...] in short there is nothing that is universally relevant.' [17] The move to abstraction in liberal thinking is, in Onora O'Neill's view, a result of the absence of homogeneous community and culture. Abstraction has been criticised for several reasons, including the view that it idealises human agency and assumes 'various superhuman capacities such as complete transitively ordered preferences, complete knowledge of the options available and their outcomes, and unwavering powers of calculation.' [17] Other criticisms of abstraction are that it omits important or material aspects of the matter at hand. But in advancing an abstract Kantian argument that rejects 'principled indifference to others' O'Neill argues for a 'realistic account of circumstances' and says defenders of abstract rights 'have to say something about the way in which obligations [...] should be allocated to individuals, office holders and institutions.' [17] According to O'Neill a right to healthcare requires counterpart duties that must be carried out by specified persons or institutions that have the relevant competency and capabilities to carry out those duties [17]. Alongside the right to healthcare, individuals and institutions can have obligations but fulfilling or discharging those obligations, necessitates individuals and institutions having adequate capabilities [17].

5. Caring relationships versus institutional care

In Gheaus' view the difference between care as work and care as relationship is mainly one of focus and if we introduce a strong emotional link as the basis of care 'we are no longer able to account for some paradigmatic cases of care-giving.' [6] Rather than looking at all types of caring to see what common skills they involve, Gheaus [6] suggests proceeding the other way round and defining the activity of caring 'via the disposition of care and its employment.' Advocating for a multi-layered understanding of care, Gheaus [6] suggests that in thinking of care work within the family or within institutions, it is not necessary to say that one type of care is more valuable than another as it is reasonable to think 'that people need various types of caring relationships during life.' Institutional care may be valuable in itself if it is a complement to the loving care one gets in intimate relationships and the caregivers are enabled to meet the needs well. On this view, care received in institutions should not be seen as a replacement to intimate relationships but as complementary and valuable in their own right:

> *'As long as it is done well, we will definitely want to call their work "care" and there is no contradiction in doing so, since the criterion for judging institutions need not be identical to those for judging people.' [6].*

Clarifying the relationship between care directed towards a particular individual involved in a personal relationship and institutional care, Gheaus argues that when the work of caring for someone is not directly motivated by the personal concern of the care-giver, care is still the moral reason behind the respective practice (for example in hospitals) but the motivation may be different and that:

> *'At its best, the work of care is concerned with wanting to meet someone's needs but in caring for strangers, the motivations may be different, and can be about money or a desire to keep jobs.' [6].*

In the following paragraphs, and in Reference to the findings of the Francis Report, and the failings of care at Mid Staffordshire NHS Foundation Trust, I hope to demonstrate how compassion, understood here as an attitude one may have towards people in general, is different to the work of care because it involves acting on behalf of others and is also limited to fewer people.

6. Mid Staffordshire NHS foundation trust: the findings of the Francis Report

In the ethics of care, needs play a central role but arriving at a precise specification of what counts as needs is particularly challenging when thinking of care-giving institutions [6]. In a series of articles in *The Lancet* the dominance of the biomedical model was identified as one of the major obstacles to giving the 'right care' to patients in acute hospitals [19]. Medics are educated according to the principles of biomedicine and value acute diseases offering the prospect of successful treatment with medical specialities demonstrating the shortest length of stay having the greatest prestige. But a growing ageing population, presenting with complex medical needs, means that older people are the major users of hospital services and often accused of 'bed blocking.' This is a term mainly used in cases where older people are deemed to be 'medically fit' but waiting for home care or alternative accommodation in a residential or nursing home setting. Nurses occupy a liminal professional space within traditional, biomedical institutions (especially hospitals) and for some, the emphasis on caring is detrimental to the profession. Paley, for example, sees care being used as a paradigm to attack the medical-scientific model of nursing that prevents its real development [4].

To be admitted to an acute hospital bed, a patient must first be suspected of having an underlying medical condition requiring treatment such as heart failure, hip replacement, pancreatitis to name but a few. In hospital settings, diagnosis and treatment are normally the responsibility of the individual physician and patients requiring medical treatment may be prescribed a range of treatments from drug therapy to surgical procedure, or both. Medical treatment is increasingly premised on evidence-based medicine (EBM), a method that typically tests traditional biomedical interventions such as drugs, devices and procedures using randomised controlled trials (RCTs) to arrive at the soundest evidence of a treatment's efficacy. Working alongside physicians treating patients, nurses use their clinical expertise and monitor a patient's progress through constant observation of vital signs such as blood pressure, bodily temperature, and administering medicines etc and assisting with basic nursing care where needed.

In the case of Mid Staffordshire NHS Foundation Trust, a distressing feature of the poor care identified by the Public Inquiry was the 'plight of patients calling for water, languishing in soiled bedding or dying neglected and confused.' [3] In her evidence to the Public Inquiry, the former Chair of Mid Staffs, Toni Brisby claimed the hospital was no different to any other and that:

> *'A reaction that I've had from quite a lot of people within the NHS, which is that actually that's the sort of thing that goes on virtually in all hospitals, and there but for the grace of God go we. Now, I'm not saying that to defend poor care, [...] but I am saying that Stafford is not a peculiar hospital.' [1].*

Concerns about nursing care highlighted by the Public Inquiry identified failures in clinical care such as completing charts, weighing patients, checking intravenous infusions, dressing wounds, and avoiding pressure sores: in several instances, patients were not helped to take their medication. Other failures involved those associated with basic nursing care including:

> *'...such matters as the supply of and help with food and drink, a timely response to call bells and buzzers, attention to the hygiene needs of patients, and respecting the dignity and privacy of patients [1].*

In an article responding to the findings of the Francis Report, Anna Smajdor dismisses the claim that a lack of compassion particularly among nursing staff is to blame for poor care and argues the root cause of poor care is a lack of time and a lack of resources in the NHS but fails to specify where the scarcity lies. On Smajdor's view compassion is not necessary when caring for patients and one can:

> *'[R]emove an appendix without caring about the person from whose body it is taken, empty a bedpan without caring about the patient who filled it, or provide food without caring about the person who will eat it.' [3].*

Conflating different types of healthcare tasks (removing an appendix versus emptying a bedpan) and the roles performed by different healthcare professionals (surgeons compared to nurses or healthcare assistants) Smajdor dismisses the need for compassion and suggests 'reminders, routines, and checklists,' can do the same job as compassion [3].

The theory of obligations advocated by Onora O'Neill suggests that those who reject 'indifference and neglect' must meet demanding standards: 'but what those standards demand is inevitably variable and selective.' [18] In the context of a ward setting, patients have many different needs, including medical and surgical needs (requiring many years of speciality training) and needs associated with basic nursing care, such as wiping bottoms, emptying bedpans, changing soiled bedsheets, or dressing seeping bed sores. The tasks associated with these latter needs are usually performed by a nurse or healthcare assistant and conflating different types of healthcare needs and the roles performed by different healthcare professionals, fails to do justice to patient suffering and of the harms inflicted through neglect, that was so much a feature of Mid Staffs NHS Foundation Trust.

Furthermore, Smajdor's analysis fails to recognise a key concept in healthcare which is the prevention of suffering and harm. The current COVID-19 pandemic has triggered debate about the right for all to access healthcare and the scarcity of resources: Mannelli [20] argues that in the present crisis 'as in other circumstances in which there is a scarcity of resources, it is unfortunately not possible to avoid harm at all. The effort is to reduce it.' At Mid Staffs such efforts were not always apparent, as patients were denied water not because water was scarce, but because ill and frail patients were neglected. Care work or 'basic' nursing as it is often referred to, is integral to the work of the nurse and if a nurse or health care assistant fails to respond to the patient who says: 'I need the toilet now' or: 'I cannot hold a knife or fork but can chew and swallow' then it is hard to see how the distressing situation of Mid Staffs can be avoided in the future.

A shift in healthcare away from one in which the doctor and patient knew each other, to healthcare provided in complex institutional settings, necessitated a refocus on medical ethics. In an effort to avoid paternalism and to protect patients, more formal relationships and procedures were instigated between the two [17]. This was achieved partly through the mechanism of *informed consent* that aimed to avoid imposing medical treatment or action on patients without being fully informed. But in O'Neill's view, there are other ethically important concerns in healthcare such as 'unnecessary surgery, clinical negligence, or unwarranted risky treatment' aside from informed consent and most ethical positions do not consider informed consent as sufficient for respecting patient autonomy:

> *'Contemporary accounts of autonomy have lost touch with their Kantian origins in which the links between autonomy and respect for persons are well argued; most reduce autonomy to some form of individual independence and show little about its ethical importance.' [21].*

Other emerging trends in healthcare include the concept of *patient-centred care* in which the ideals of independence and self-care are promulgated. The phrase 'patient-centred care' originated in the United States but has gained prominence in the UK [22]. Linked to the shift in healthcare from paternalism to autonomy, patient-centre care (or person-centred care as it is more frequently known) is intended to represent the shifting of power and control from the healthcare provider or practitioner, to the patient. With patient autonomy taking precedence over paternalism or 'best interests' Jonathan Evans argues that advocating independence for patients who are dependent upon others for help may be worthless without the necessary power needed to secure the care and attention they require [23].

Up until the late 1970s, Anne Bradshaw [4] notes that nearly every survey into nursing care contained unsolicited comments on the kindness and helpfulness of nursing staff but that patient perceptions changed in the 1990s. Under Tony Blair's government, attempts at measuring compassion using patient surveys, was proposed by then Secretary of State for Health, Alan Johnson who said patients had a right to be treated with 'dignity, respect and compassion' – a move that was supported by the Royal College of Nursing [4]. Dismissive of such moves Anne Bradshaw [4] claims it is inherently false to measure and reward the appearance of compassionate care (such as encouraging nurses in the art of smiling or the saying of warm words) for the purposes of data collection. For Anna Smajdor the 'insidious' need to measure all we touch, including compassion, is part of a broader trend that is in awe of the evidence-based structure of our health service:

> *'Some – perhaps much – of the suffering experienced by patients and healthcare professionals in today's healthcare systems, is the result of a clash between incompatible values. On one side, there is a scientistic ideology which holds that everything which is meaningful must be measurable and controllable. On the other, there is the conviction that some of the most valuable things in life are intrinsically so; [...] It is not compassion per se, that is at issue here, but a far broader and more insidious need to measure all we touch.' [3].*

Following concerns about variability in medical practice and rising costs, efforts were made to make such systems more quantifiable by introducing a range of initiatives including compulsory clinical audit, quality assurance (QA) and evidence-based medicine (EBM.) [5, 19] The Department of Health in Britain in 1994 embraced the notion of knowledge-based medicine and assumed that science could identify non-effective treatments or procedures thereby creating uniformity in the delivery of services to various patients [5]. This 'scientistic ideology' according to Smajdor has contributed to much of the suffering experienced by today's patients and healthcare staff alike and has stifled our ability to care. But given evidence-based medicine is primarily concerned with the effectiveness of drugs, devices and procedures and not tasks associated with assisting patients on and off the commode, or helping with food and drink, Smajdor's claim is somewhat puzzling. This is particularly so as such tasks are typically exempt from the scientific impulses driving evidence-based medicine. And, as effectiveness and outcomes represent values and not scientific universals [5], outcomes that may be seen as good from the doctor's perspective (the patient is medically fit to go home) are not necessarily good from the patient's perspective (the medically fit patient awaiting discharge is left to languish in soiled bedding).

Furthermore, Smajdor's suggestion that 'reminders, routines, and checklists,' can do the job of compassion may only further inhibit 'human interaction and thinking, lead[ing] to an increasingly rationalised world.' [24] Such tools may be superfluous to the basic task at hand and potentially damaging to its performance. Plus, the utility of such tools when compared to the potent 'reminder' of a patient calling out for help, or the pungency of a bedpan that needs emptying remains questionable. The NHS is a much criticised and much loved organisation that some argue has taken the place of religion [25]. Misdiagnosing all that ails the NHS, coupled with gratuitous sentimentality can only prove fatal to its proper functioning in the long run.

7. Care of the self

As the capacity to care and respond to an individual's needs is one of the defining characteristics of being human, losing our ability to care can be harmful. Caring *about* something constitutes a need in itself as it can bring meaning into our lives because life perceived to be meaningless can lead to depression [6]. In the Francis Report there were many accounts in which healthcare professionals employed at Mid Staffs, expressed their distress and feelings of depression and helplessness at finding their concerns dismissed [3]. In looking to identify the causes of the Trust's failings, the Public Inquiry investigated the effectiveness of the Trust's whistleblowing policies that were 'intended to empower employees to raise concerns.' From its review of the Trust's actions in the case of an A&E nurse, the Inquiry concluded Mid Staffs did not follow its stated whistleblowing policy of supporting and protecting those who raised concerns:

> *'Ms Donnelly was offered no adequate support. She had to endure harassment from colleagues and eventually left for other employment. Clearly such treatment was likely to deter others from following her example and she was aware of colleagues on whom this had an effect.' [1].*

It was also revealed that Ms. Donnelly was failed by her professional organisation, the Royal College of Nursing (RCN). The Inquiry also noted that doctors who sought to raise concerns fared little better than nurses. In evidence given to the Inquiry by Dr. Pradip Singh, the doctor admitted he was not 'brave enough.' He told the Inquiry raising concerns would have had a detrimental effect on his health and he also feared losing his job: 'I would have then ended up becoming either a stroke or a heart attack, and being on the road.' [1].

One difficulty with care-giving institutions is the number of people needing care at any one time. This can have implications for the quality of care and can lead to the overburdening of care-givers. Unlike a utilitarian or Kantian perspective on morality, which states that one must not place the requirements of self above the requirements of others, the ethics of care makes allowances for the better care of the self and recognises the moral importance of ensuring one's own needs are met [6]. This is a view also articulated by Mary Wollstonecraft; writing in the eighteenth century on women's human rights, she advised her female readership - often deprived of opportunities for self-development - that their 'first duty is to themselves as rational creatures.' [26] In conversation with a former psychiatric nurse, she described how she was distressed at the manner in which patients were treated on the ward she worked, and explained she liked to treat patients in the same way she would treat a relative, or how she herself would like to be treated. On referral to a counsellor (provided by her employer) she was told it was unrealistic to expect her colleagues to care for patients in the same way she did, and having such expectations, would only be detrimental to her own well-being. She has since left the profession and now does odd jobs to earn a living.

At its best, the work of care is concerned with wanting to meet someone's needs and with compassion one of the six values of the NHS Constitution, staff are expected to demonstrate compassion and kindness as part of their work. Staff placed under unremitting pressure can however become estranged from each other and those 'coerced by circumstances become coercers.' [11] Michael West writing for the NHS Leadership Academy maintains NHS leaders need to embody the virtue of compassion because in an environment that is:

> *'directive, controlling, punitive, threatening or uncaring, [...] compassion dries up and [...] bullying becomes dominant [27].*

And in the context of care-giving institutions, care-givers need to be able to rely on people and institutional structures to support them because:

> *'For those invested with institutional power, though the responses will be different from the requirements that apply to non-institutional interaction there is no reason to believe that a citizen who wishes to live in a caring society would not have any more reason to tolerate institutional abuse than the one who wishes to live in a just society.' [6].*

The NHS Constitution lists compassion as one of six values that is central to the work of the organisation that is realised through the expression of humanity and kindness to both patients and fellow staff members. In their book *On Kindness*[11], Phillips and Taylor write that the pleasures of kindness are fundamental to our sense of well-being, a view also shared by the Mental Health Foundation in the UK that chose kindness as its 2020 theme:

> *'[kindness] Has the singular ability to unlock our shared humanity. Kindness strengthens relationships, develops communities and deepens solidarity. It is a cornerstone of our individual and collective health.' [22].*

But unkindness is now the norm in our society, according to Philips and Taylor and overcoming our current attitude towards kindness, requires a form of 'ordinary, unsentimental kindness' because:

> *'Real kindness is not a magic trick, a conjuring away of every hateful or aggressive impulse in favour of a selfless dedication to others. It is an opening up to others that [...] enlarges us and so gratifies our profoundly social natures.' [11].*

Following Freud and Winnicott, both Phillips and Taylor argue that sentimentality and nostalgia, and not hatred, are the enemies of kindness with too much kindness seen as a saboteur of fully formed independence. Gaita [28] agrees we often struggle against a disposition to sentimentality that prevents us from seeing things as they are, rather than as they appear:

> *'When concepts such as sentimentality, pathos are causes of the false, they are psychological states that can cause thought to go astray more or less as tiredness, drunkenness, fearfulness or recklessness can.' [28].*

Seeing things as they are and not as they appear, requires a form of understanding in which head and heart are inseparably combined and says Gaita 'is neither a Kantian nor a Humean thought, but one which acknowledges what is important to both of these traditional oppositions.' [28].

8. Conclusion

Drawing on the parable of the Good Samaritan and the work of Anca Gheaus and the ethics of care, this paper has explored how the concepts of compassion and care differ. The distinctiveness of care according to Gheaus lies in the desire to actively help when possible and is a more intense and engaged activity compared to compassion. In the parable of the Good Samaritan, the difference between compassion and care is best illustrated by the differing levels of involvement of the Samaritan who had the resources to pay someone to do the caring, and the innkeeper who agreed to care for the victim in return for money. As the work of care is concerned with wanting to meet someone's needs, the motives for caring may be different depending on context, and for those in care-giving institutions, their relationship with patients remains largely instrumental and can be about money and staying employed. In extrapolating from the parable of the Good Samaritan to contemporary healthcare, the paper follows O'Neill and her arguments on the utility of abstraction and the need to take a realistic account of circumstances.

Bureaucratisation and rationalisation have been blamed for usurping the value of care with the traditional approach to care as 'my brother's keeper' falling apart in the face of such efforts. With utilitarianism now considered to be a major influencer in the practice of medicine, its influence has also seeped into UK nursing. Unlike compassion - here understood as an attitude one may have towards people in general - the work of care is concerned with wanting to meet someone's needs. The ethics of care is opposed to a utilitarian approach to care and argues for care to retain its special moral significance, a vivid sense of particular situations and concrete individuals is necessary.

In the context of healthcare, meeting the needs of concrete individuals, remains a challenge to the NHS. As care is a more intense and engaged activity according to the ethics of care, defining what counts as needs, and determining how such needs are met, particularly among patients who are most vulnerable, requires a form of thought in which head and heart are intertwined. The dangers to a healthcare system that encourages staff to rely on 'reminders, routines and checklists' rather than thinking and human interaction along with ordinary, unsentimental, everyday kindness, can only give cause for concern, and may do little to avoid the harms previously experienced by patients and staff at the Mid Staffordshire NHS Foundation Trust.

Author details

Una P. Canning
London, UK

*Address all correspondence to: u.canning1@gmail.com

References

[1] *Report of the Mid Staffordshire NHS Foundation Trust Public Inquiry, Vol. 1; Analysis of Evidence and Lessons Learned.* https://assessets.publishing.service.gov.uk/government/uploads/system/uploads/attachment_data/file/279115/0898_i.pdf

[2] NHS foundation trusts | The King's Fund. Accessed April 22, 2021. https://www.kingsfund.org.uk/publications/articles/nhs-foundation-trusts

[3] Smajdor A. Reification and compassion in medicine: a tale of two systems. *Clin Ethics*. 2013;8(4):111-118

[4] Bradshaw A. Measuring care and compassion: the McDonalised nurse? *J Med Ethics*. 2009;35:465-468

[5] Dew K. *The Cult and Science of Public Health: A Sociological Investigation*. Berghhan Books; 2012. doi:10.5860/choice.50-0349

[6] Gheaus A. Care and Justice: why they cannot go together all the way. PhD thesis beforehand. Published online 2005. https://scholar.google.co.uk/citations?view_op=view_citations&hl=en&user=dBtDbXIAAAAJ&citation_for_view=dBtDbXIAAAAJ:d1gkVwhDpIOC

[7] Gheaus A. Love, hate and moral inclusion. In: Carlisle, J. Carter, J.C. and Whistler D, eds. *Moral Powers, Fragile Beliefs*. Continuum Publishing; 2011:29-52

[8] The NHS Constitutional Values Hub. https://www.hee.nhs.uk/about/our-values/nhs-constitutional-values-hub-0

[9] Tronto J. *Moral Boundaries: A Political Argument for an Ethic of Care.* Routledge; 1993

[10] Arbuckle G. *Humanizing Healthcare*. Jessica Kingsley; 2012. doi:10.4324/9781351266369-17

[11] Phillips A and Taylor B. *On Kindness*. First edit. Hamish Hamiliton; 2009

[12] Allen D. The legacy of Project 2000. *Nurs Stand*. 2009;23(34):18-21. doi: 10.7748/ns.23.34.18.s24

[13] *Old Testament, Leviticus*

[14] Ruddick S. *Maternal Thinking. Towards a Politics of Peace.* Beacon Press; 1989

[15] Kittay EF. At the margins of moral personhood. *Ethics*. 2005;116:100-131

[16] Gaita R. *A Common Humanity. Thinking about Love and Truth and Knowledge*. Routledge; 2000

[17] O'Neill O. *Justice Across Boundaries: Whose Obligations?* 3rd ed. Cambridge University Press; 2017. doi:10.1017/cbo9781316337103

[18] O'Neill O. *Towards Justice and Virtue: A Constructive Account of Practical Reasoning.* Cambridge University Press; 1996

[19] Valles SA. *Philosophy of Population Health: Philosophy for a New Public Health Era.* Kindle edn. Routledge; 2018

[20] Mannelli C. Whose life to save? Scarce resources in the COVID-19 outbreak. *J Med Ethics*. 2020;(46):364-366

[21] O'Neill O. Some limits of informed consent. *J Med Ethics*. 2017;29:103-106. doi:10.4324/9781315240046-9

[22] The Health Foundation. *Person-Centred Care Made Simple: What Everyone Should Know about Person-Centred Care*. October 2014

[23] Evans J. Person-Centred Care and Culture Change. Caring for the Ages. 01 Aug 2017;18(8):6

[24] Ritzer G. *The McDonaldization of Society*. Pine Forge Press; 1992

[25] Let's be frank about the NHS | The King's Fund. Accessed April 23, 2021. https://www.kingsfund.org.uk/blog/2018/03/lets-be-frank-about-nhs

[26] Botting Hunt E. *Wollstonecraft, Mill and Women's Human Rights*. Yale University Press; 2016

[27] West M. Compassion is the core NHS value - not bullying. https://www.leadershipacademy.nhs.uk/2016/09/28/

[28] Gaita R. Morality, metaphysics and religion. In: Carlisle, J. Carter, J.C. and Whistler D, eds. *Moral Powers, Fragile Beliefs*. Continuum Publishing; 2011:3-28

Section 2

Inequalities and Disparities in Healthcare

Chapter 4

The Risk of Inequality in Italian Healthcare Due to Covid-19

Carlo Ciardo

Abstract

The Italian National Health Service, characterized by the principles of universality, equality and fairness, has undergone changes over the years that have involved these essential characteristics. The decrease in financial resources was the first element that touched the Italian health organization. The spread of Covid-19 has attacked the balance of healthcare in Italy and put the equality of the entire care system at risk. The reform of the Italian health system, especially through the correct use of European financial resources, is the real test for the Italian health system of the future. It can be a moment of relaunch or the certification of a decline that jeopardizes constitutional rights.

Keywords: healthcare, inequality, Italy, covid-19

1. Introduction

1.1 The fundamental principles of the Italian Health Service and the criticalities of public health expenditure

The Italian National Health Service, characterized by the principles of universality, equality and equity, has suffered over the years an erosion of the aforementioned pillars also due to the systematic decreases in funding. In addition, the balance of the entire social welfare system is today subjected to an unexpected and very significant attack caused by the health emergency dictated by COVID-19.

The Italian model, created by Law n. 833/1978, was invoked evidently to the British experience of the original N.H.S. [1], while being distinguished by an accentuation of the intervention public. Wanting to summarize the criteria of the law establishing the NHS, 5 key points can be identified: a) "universality" of the provision of assistance services, b) "uniqueness" of the management of health facilities and hospitals by the USL (Local Health Units), the real engine of the reform; c) "equality" in carrying out therapeutic treatments; d) planning of assistance objectives and resources financial.

The 1978 reform implemented the constitutional provision of art. 32 of the Constitution, preparing a health network suitable for an organic and global protection of health. Just the art. 1 of the aforementioned law openly recalls the constitutional text, defining the protection of health «as fundamental right of the individual and collective interest» (art. 1, paragraph 1) and in paragraph 2, following the provisions of art. 32 of the Italian Constitution: «the protection of physical and mental health must take place in respect for the dignity and freedom of the human person». Strictly consequential with respect to this regulatory provision, it is the direct intervention of the public authorities in the organization and in the provision of assistance itself.

In this regard, it is necessary to highlight that the definition of health protection does not only concern the absence of disease or disability. These references are now reductive and simplistic, so much so that the Organization World Health Organization has stated that health must be done to mean "a state of complete physical, mental and social well-being [...] Condition of harmonic functional and physical balance and psychic organism dynamically integrated into its environment natural and social" [2]. Thus, the notion of health cannot yet be linked to the absence of infirmity, but has undergone a significant expansion of both a biological and social nature, alongside integrity physics of a static nature, also a dynamic element e relational [3].

In the light of the constitutional and normative arguments already mentioned, the solidity of the founding principles of the nal Health Service outlined since 1978 is of primary importance. Subsequently - already represent a substantial modification of the original structure on which the epidemic from Covid-19 had a decisive impact.

With reference to financial resources, it should be noted that the crisis of 2007–2008 had already triggered a spiral of decrease in economic allocations also through the provision of cost containment measures (for example: increase in the sharing of the so-called ticket and blocking of turnover).

The aforementioned measures have achieved, in the short term, the objectives set, so much so that in 2012 there was a beginning of a decrease in public health expenditure. This reflection, however, must be completed with a broader examination. In order to fully understand the overall picture, it is necessary to highlight that healthcare expenditure depends on multiple components: the aging of the population, epidemiological change, scientific progress and the change in information asymmetries between doctor and patient. The variation of each of these factors determines a change in the demand for services and therefore has an impact on the Health Service and its economic needs.

1.1.1 Factors affecting the organization and financing of health care

The first element to be evaluated is the demographic one. It is undeniable that there has been an increase in life expectancy. In Italy, from 1960 to today, life expectancy at birth has increased by 12.8 years for women and by 13.6 years for men, so much so that the average age of women is 85.2 years and for men it is 80.8 years. The national average is 83 years compared to the 80 years of the OECD average [4].

These data, together with the slowdown in the birth rate, have produced an increase in the elderly population, to which new welfare needs are linked.

The exponential increase in chronic diseases (for example: cardiovascular diseases, cancer, diabetes and dementia) - those most present in the elderly population - has a significant impact on both the organizational structure of the National Health Service and the economic one. Suffice it to say that in Europe in 2016 the treatment of cardiovascular diseases alone cost 192 billion euros, much higher than the expenditure necessary for the diseases that most afflict the younger population [5].

Another factor that affects the health organization and its financing is that linked to scientific progress and technological innovation. This element does not depend on the choices of each individual country, but on investments in research and development by manufacturing companies (especially pharmaceutical companies). The public resources made available in this field have the effect of conditioning the degree of access to knowledge and new technologies. In summary, the availability of about 14,000 medical device patents each year provides answers to health problems and creates, on the other hand, a new perceived need and, consequently, a new demand for assistance with undeniable repercussions on health care costs.

Think that the growth of technological innovation is much higher than the growth rate of the economy, so much so that between 2008 and 2012 while Italy recorded a decrease in GDP equal to 2.3%, in Europe 1 patent for medical devices was registered every 38 minutes [6], proving the fact that the National Health Service cannot keep up with the technical and scientific developments.

The last element of this picture of factors is represented by the variation of information asymmetries between doctors and patients dictated by the ease of acquiring information through the mass media and, above all, through the internet. If this increases the levels of knowledge of citizens, on the other hand, information bulimia also increases the risk of inappropriate requests, up to the so-called defensive medicine, a prescriptive modality that does not safeguard the patient's health, but removes the risk of professional liability, increasing inappropriate spending.

All the aforementioned affect the National Health Service and determine an ever-increasing and different demand for services, moreover in an emergency context dictated by the spread of COVID-19, which requires new financial flows and imposes a radical organizational change.

2. The right to health protection "conditioned" by financial and organizational resources in the light of constitutional jurisprudence

The entire issue of the health organization must also be read in the light of constitutional jurisprudence.

The Italian Constitutional Court, in implementation of art. 32 of the Constitution [7], has ruled that the right to health protection, understood as a request for services and benefits, is "primary and fundamental" and requires "full and comprehensive protection" [8].

This interpretative orientation changed, however, in the 1990s, when the need to contain public spending was imposed [9]. In the last decade of the last millennium, in fact, the Italian public debt had come close to 100% of GDP, while the public deficit was 11%. Especially since 1992 the situation has appeared out of control, so much so that the successive Governments (the first in this sense was the one chaired by Prof. Giuliano Amato) have begun a new phase of fiscal consolidation, with inevitable effects also on health expenditure [10].

The sentence of the Constitutional Court no. 455/1990 with which, while reiterating that the right to obtain health services must be guaranteed to every individual, he stated that this right is conditioned by the legislative choices of balancing the protection of health with other constitutionally protected interests: "taking into account of the objective limits that the legislator himself encounters in his work of implementation in relation to the organizational and financial resources available at the moment" [11]. For the first time it was established that the right to health protection had to deal with economic and organizational possibilities [12].

The subsequent evolution of constitutional jurisprudence has confirmed the need to achieve a balance between the right to benefits and economic and organizational limits, but has also provided for an essential core of the right to the protection of health that can never be undermined [13]: "in the balance of constitutional values operated by the legislator had an absolutely preponderant weight, such as to compress the essential core of the right to health connected to the inviolable dignity of the human person, we would be faced with a macroscopically unreasonable exercise of legislative discretion" [14].

Constitutional jurisprudence relating to the patient's freedom of choice is also of considerable importance. Precisely the need to take into account the limited economic resources and the need for general health planning led the Italian Constitutional Court to establish that the patient's freedom of choice of the health facility at which to carry out a diagnostic examination or a medical intervention cannot be considered as an absolute right because of the incompressible economic needs [15].

The above is a confirmation, from the wall of the Italian Constitutional Court, of the policies to contain health costs, albeit within the limits of a reasonable balance with the various constitutionally relevant interests.

3. The evolutionary lines of the National Health Service and the legislative choices of the Italian Regions

Looking at the evolution of public policies in the health field, one must take into account the multiple choices made by the Italian Regions in the organization of Regional Health Services.

In most cases there has been a progressive unification of Local Health Authorities (ASL) and Hospitals (AO). One of the first Regions to make this choice was the Marche Region with the creation in 2003 of the Single Regional Health Authority (ASUR). This example was followed by the Abruzzo Region, with only 4 ASLs, and by Umbria, which passed to 2 ASLs.

Further examples can be given on the choice of reduction. Puglia already after the health reform of 1992 had gone from 55 Local Health Units to 12 ASLs, but in 2007 it decided to carry out another merger until it reached only 6 ASLs.

In 2015, the Tuscany Region reduced the ASLs from 12 to 3, and 4 university companies were added to these. In 2014, the Emilia Romagna Region created a single ASL (for the territories of Forlì, Cesena, Ravenna and Rimini) with over 1 million inhabitants, 12,000 employees and a budget of 2.2 billion euros.

In 2015, on the other hand, the Lombardy Region replaced the ASLs with 8 Health Protection Agencies and 25 Territorial Social and Health Companies.

Analyzing the different choices of the Italian Regions, in an effort of synthesis it can be said that the mergers were designed to create economies of scale, to simplify administrative procedures, to eliminate duplications. It should also be noted that these choices, in the medium term, can also cause inefficiencies linked to the excessive size of the ASLs and difficulties in creating middle management.

The above must be completed with an assessment of the evolution of the hospital organization.

The Law Decree n. 95 of 6 July 2012 provided for a reduction in hospital beds, so much so that a maximum value of 3.7 beds per 1,000 inhabitants was reached, compared to the previous value of 4 beds per 1,000 inhabitants. To this was also added an evaluation of small public hospitals in order to dismiss these structures and favor home care.

The process of reducing the number of beds and reorganizing the network of small hospitals was implemented with the decree of the Ministry of Health no. 70 of 2 April 2015 which defined the qualitative, structural, technological and quantitative standards. This legislative choice was aimed at reducing health care costs, reorganizing the hospital network and moving health care for many chronic diseases outside hospitals to put it at the expense of "territorial assistance".

In reality, however, this reorganization made the mistake of having accelerated the rationalization of hospitals by the Italian Regions, but it was not as effective as regards the creation of a new territorial assistance which, therefore, remained incomplete in many parts of the national territory.

4. Effects of public policies on the National Health Service

After the economic crisis of 2008 there was a gradual reduction in the financing of the National Health Service. In the period between 2009 and 2014, Italian public health expenditure increased each year by 0.7%, while from 2003 to 2008 the growth in health expenditure was equal to 6%.

These policies to contain national health expenditure and the choices of the Italian Regions to reduce their budget, have had undeniable positive effects both on the national budget and that of many Regions that were in financial crisis (think of Regions such as Lazio, Campania and Sicily). At the same time, however, these choices have had an impact on health care.

The first effect of the reduction of economic resources for the National Health Service was the reduction of some assistance services. This aspect had a fundamental impact on citizens as those who had the financial availability were able to turn to private health facilities, while those who could not afford this expense independently had to give up temporarily.

This has also led to an increase in patient waiting lists at public health facilities. The "Annual report on hospitalization activities" of the Italian Ministry of Health (2019) contains some significant data. For example, in 2017 the average expectations were over 27 days for breast cancer, 53 days for prostate cancer, 119 days for tonsillectomy, 90 days for inguinal hernia [16].

In Italy 40% of specialist health visits, 49% of rehabilitation services and 23% of diagnostic tests are paid directly by citizens [17].

In addition, the policies to contain health costs also had a direct influence on medical personnel. In Italy there is no shortage of doctors, so much so that every year there are about 9,000 graduates in medicine and surgery, but only 6,000 graduates each year can receive a specialist scholarship. This means that every year in Italy about 2,000 doctors live in the uncertainty of their professional future, despite the enormous need for new doctors by the National Health Service.

To confirm what has been said, the fact that in the Veneto Region there is a deficit of 400 doctors, in the Piedmont Region and in the Puglia Region there is a need for about 300 doctors. If we look up to 2025, a national shortage of about 16,700 medical specialists is expected. An incredible paradox [18].

5. A change of legislative step and the possible repercussions deriving from the COVID19 emergency

It must be borne in mind that in recent years there has been a change in the choices of health policies. The resources allocated to the National Health Service have been slowly increased, although still largely insufficient. In the 2019 Economic and Financial Document, health expenditure for 2018 was estimated at approximately 115.4 billion euros, recording a growth of 1.6% compared to 2017.

This precarious balance has, however, been completely upset by the emergency caused by the spread of COVID-19 which has imposed a campaign of staffing and a sudden technological and infrastructural investment of which the exact extent is not yet known. In this regard, it must be taken into account that the first legislative intervention approved in 2020, in full epidemiological emergency, was the "Cura Italia" Law Decree, which immediately provided for the following emergency measures: additional funding for incentives to healthcare personnel (Article 1); hiring by the Minister of Health of 40 medical executives, 18 veterinary executives and 29 non-executive personnel, allocating over 5 million euros for 2020, 6,790,659.00 euros for 2021 and 2022 and almost 1,700 € .000.00 for 2023 (Article 2); possibility for the Regions to purchase medical equipment also from private health facilities (art. 3); an increase in the financing of healthcare for an amount of almost 2 million euros (Article 17).

This was only the first step in a series of economic increases which during 2020 (and also at the beginning of 2021) were expected to address a condition that has upset not only Italian society, but above all the entire National Health Service.

All this must also be taken into account the impact that the national and global financial crisis that was triggered by the pandemic, moreover on the Italian economy, which even before the health emergency had shown signs of weakness [19]. One of the first reports by the Moody's agency supported a possible recession in the Italian economy with a reduction in GDP of over 10%. For this reason, what must be watched carefully is the behavior of both the European Union and that of the ECB.

6. Is Italian healthcare likely to be unequal?

The evolutionary framework of the interventions of economic and organizational policies in the health field gives us a plurality of indices from which multiple inequalities emerge, so much so as to touch the foundations of the constitutional right to the protection of health.

The first imbalance is between Northern Italy and Southern Italy. Already today in the Northern Regions the life expectancy for men and women is respectively 81.2 years and 85.6 years, while in the South the life expectancy for men and women is 79.8 and 84.1 years. Years.

An index of inequality is represented by healthcare mobility between Regions, so much so that in 2017 88% of Nitalians who moved to be treated were hospitalized in Lombardy, Emilia Romagna, Veneto, while 77% of citizens who moved to try to be treated came from the Regions of Central and Southern Italy (Puglia, Sicily, Lazio, Calabria and Campania).

A further sign of inequality is linked to the renunciation of care by the poorest sections of the population. 43% of Italians with serious economic difficulties declare that they are not in good health, while only 23% of those who do not have economic difficulties believe that they are not in good health [Italian Higher Institute of Health, "Report on inequalities", 2019]. It must also be assessed that 55% of people with a low level of education report that they are in poor health (compared to 20% of Italian graduates) and 12% say they have depressive symptoms (compared to 4% of graduates).

This context of inequalities is even more relevant if we consider that over 11 million Italians in 2017 gave up on treatment for economic reasons and 7 million Italians went into debt to take care of themselves.

This fragile health care landscape is also characterized by significant paradoxes, the first of which is linked to the ineffectiveness of care services. It has been estimated that 19% of public spending, 40% of private household spending and 50% of the expenditure incurred by insurance policies, is used for diagnosis or for inappropriate or ineffective health services.

All this represents the photograph of a National Health Service that was already present before the COVID-19 emergency was characterized by gaps and contradictions and that, with absolute probability (if not certain), the spread of the virus will only exacerbate, especially with regard to of the poorest faces of the population, those socially more fragile and, in particular, towards the elderly.

We are in the presence of welfare, territorial and access to care inequalities that risk completely undermining not only the foundations of the National Health Service, but also those of the constitutionally protected right to health protection.

Furthermore, still on the subject of inequalities, the annual reports of the World Bank, the International Monetary Fund and the OECD, which constantly provide data on the increase in the concentration of wealth, cannot fail to have reinterpreted on the welfare state and, consequently, on the gap between the assistance potentials provided by technical-scientific progress and the actual levels of assistance provided.

From the above it follows that the picture of the evolution of Italian healthcare, also in the light of the contingencies that the current health emergency, opens the way to pressing fundamental doubts about the effectiveness of the principles of universality, equality and equity that inspired the creation of our SSN and which today risk being only chimeras.

Faced with this complex picture, it is necessary to think not only about how to convert entire healthcare companies into hospitals dedicated to the care of patients suffering from Covid-19, moving patients with other diseases to other small hospitals, but it is also urgent to already start planning a future in which there are integrated assistance systems, with large hospitals that are centers in which investments are concentrated and specialization is developed, connected in telemedicine with small hospitals.

In this new vision, one cannot imagine addressing the issue of small hospitals only with a view to reducing costs, but it is necessary to plan their use in a perspective of continuity of care and strengthening of territorial assistance.

On the side of medical personnel, the time has now come to face the season of the shortage of doctors and nurses with determination, nor can we think that this problem can be solved exclusively with urgent procedures for the recruitment of usable staff to fight the pandemic. Instead, it is necessary to design new care models and launch real training programs for the phase following Covid-19. The contingent criticality must be transformed into an opportunity to strengthen a poor health system, first of all, from the point of view of the staff. It will be possible to plan the reconversion of services by leveraging the energies and skills of young doctors to be included in the National Health Service.

Furthermore, it would be short-sighted to limit ourselves to asking for economic policies to increase public spending to be allocated to the health service (back to a percentage of about 7% of GDP), nor would it be satisfactory to limit ourselves to hoping for investments for the adaptation of structures. Hospital. Instead, it is necessary, right from the start, to train young doctors and nurses, as well as new professional figures capable of bringing new services.

Finally, any reorganization of the National Health Service would be empty if the problem were not raised (and the solution started) of strengthening local health care, because only through this simple (but essential) Copernican revolution will we be able to address both the needs imposed by the pandemic and the needs that health care has to face every day in order to provide answers to the many health protection questions raised by the population.

Author details

Carlo Ciardo
University of Salento, Italy

*Address all correspondence to: carlociardo@hotmail.com

References

[1] G. Cazzola, *La sanità liberata*, Il Mulino, Bologna 1994, p. 23; F. Longo, E. Vendramini, *La nuova riforma NHS: è morto il foundholding, viva il foundholding*, in Mecosan, 1999, 2.

[2] This WHO deficit was reported by G. Cilione, *Diritto Sanitario*, Maggioli, Rimini 2005, p. 15. On the right to the protection of health as a balance between "soma" and "psyche" see also the jurisprudence of the Italian Constitutional Court, sentences nn. 161/1985, 215/1987, 167/1999, 181/2002.

[3] M. Luciani, *Salute*, *Diritto alla salute (dir. Cost.)*, cit., p. 6. Cfr. M.S. Giannini, *"Ambiente": saggio sui diversi suoi aspetti giuridici*, in *Riv. trim. Dir. pubbl.*, 1973, p. 23 ss.; B. Caravita, *La tutela dell'ambiente nel diritto costituzionale*, in Aa.Vv., *Diritto pubblico dell'ambiente*, Cedam, Padova 1996, p. 43 ss.

[4] OECD, 2019, www.oecd.org 17/03/2020.

[5] The European House-Ambrosetti, *Meridiano Sanità - Rapporto 2018*, https://www.sanita24.ilsole24ore.com 12/01/2020.

[6] EUCOMED 2011 in Ministero della Salute – CERGAS, *Report on the expenditure recorded by the public health facilities of the NHS for the purchase of medical devices*, Roma, dicembre 2011, www.salute.gov.it 17/03/2020. Sul punto cfr. R. TARRICONE, *Innovation and competitiveness in regulated industrial systems. Medical device companies,* Milano, 2010.

[7] D. MORANA, *La salute nella Costituzione italiana, profili sistematici*, Milano, 2002. B. CARAVITA, *La disciplina costituzionale della salute*, in *Diritto e società*, 1984, p. 31; M. LUCIANI, *Diritto alla salute (dir. Cost.)*, in *Enciclopedia Giuridica*, Roma, 1989, 21, p. 5. A. PACE, *Problematica delle libertà costituzionali, parte generale*, Padova, 1990, p. 44.

[8] Corte Costituzionale n. 992/1988. Corte Costituzionale nn. 103/1977, 88/1979, 184/1986, 559/1987, 992/1988, 1011/1988, 298/1990, 184/1996. B. PEZZINI, *Principi costituzionali e politica della sanità: il contributo della giurisprudenza costituzionale alla definizione del diritto sociale alla salute*, in C.E. GALLO, B. PEZZINI (a cura di), *Profili attuali del diritto alla salute*, Milano, 1998, p. 19. A. COSENTINO, *Diritto alla tutela della salute e terapie alternative, le scelte dell'amministrazione sanitaria ed il controllo dei giudici: riflessioni in margine al caso Di Bella*, in E. BRUTI LIBERATI (a cura di), *Il "caso Dibella" nella giurisprudenza civile*, Milano 2003, p. 18.

[9] A. ANZON, *L'altra "faccia" del diritto alla salute*, in *Giurisprudenza costituzionale*, 1, 1979, p. 656; R. FERRARA, *Il diritto alla salute: i principi costituzionali*, in S. RODOTÀ, P. ZATTI (diretto da), *Trattato di biodiritto*, Milano, 2010, 4, p. 21.

[10] F. CERNIGLIA, *La spesa pubblica in Italia. Articolazioni, dinamica c un confronto con gli altri Paesi*, Università cattolica del sacro Cuore - Vita e Pensiero, 2005 pp. 34-47.

[11] A. SPADARO, *I diritti sociali di fronte alla crisi (necessità di un nuovo "modello sociale europeo": più sobrio, solidale e sostenibile)*, in *Rivista AIC*, 4, 2011, pp. 1 ss.; D. MORANA, *I diritti a prestazione in tempo di crisi: istruzione e salute al vaglio dell'effettività*, in *Rivista AIC*, 4, 2013, pp. 1 ss.; E. CAVASINO, G. SCALA, G. VERDE (a cura di), *I diritti sociali dal riconoscimento alla garanzia*, Napoli, 2013; S. GAMBINO (a cura di), *Diritti sociali e crisi economica. Problemi e pro-spettive*, Torino, 2015. R. NANIA, *Il diritto alla salute tra attuazione e*

sostenibilità, in M. SESTA (a cura di), *Attuazione e sostenibilità del diritto alla salute*, Roma, 2013, pp. 29-44.

[12] R. FERRARA, *Tutela della salute: prestazioni amministrative e posizioni soggettive*, in M. ANDREIS (a cura di), *La tutela della salute tra tecnica e potere amministrativo*, Milano 2006, p. 44.

[13] B. PEZZINI, *Principi costituzionali e politica della sanità: il contributo della giurisprudenza costituzionale alla definizione del diritto sociale alla salute*, in C.E. GALLO, B. PEZZINI (a cura di), *Profili attuali del diritto alla salute*, Milano, 1998, pp. 28 ss.

[14] Corte Costituzionale n. 304/1994. In tal senso, *ex multis* Corte Costituzionale nn. 309/1999, 432/2005, 269/2010, 61/2011.

[15] Corte Costituzionale n. 111/2005; in tal senso altresì cfr. Corte Costituzionale n. 94/2009.

[16] Ufficio parlamentare di Bilancio, *Lo stato della Sanità in Italia,* Focus n. 6, 02.12.2019.

[17] CENSIS-RBM, *9 Rapporto Rbm-Censis 2019,* www.welfareday.it *17/03/2020.*

[18] ANAAO Assomed, 2019.

[19] ISTAT www.istat.it 11.01.2020.

Chapter 5

Radical Bioethics: Difference, Disability, and Desiderata

Mary Jo Iozzio

Abstract

With diatoms—globally abundant single cell algae—as both a model and an extreme example of diversity among a single species, Radical Bioethics examines narrow constructions of human diversity as a failure of imagination and a refusal to recognize disability as another instance of difference. Along with other disciplines, bioethics has been slow to consider its biases, inherited from a history of social constructions, against people with disability. Both desire and desiderata offer an alternative to harms committed against people with disability in matters relating to initiatives that foster their inclusion as critical participants in and rightful recipients of the commonweal.

Keywords: bioethics, desire, desiderata, disability, diversity, historical consciousness, theological anthropology, United Nations Convention on the Rights of Persons with Disability, World Health Organization

1. Introduction

Knowingly or not, disability in the human community is a global reality about which too few register a thought concerning its prevalence. As a result of this unrecognizability, this failure to register, too few people without immediate experience of or any regular encounter with persons with disability remain unconcerned with this largest and most diverse minority of people across the globe (by the World Health Organization estimates, at least 15% of Earth's human population). Moreover, the likelihood of able-bodied/able-minded persons joining this minority increases with age if not by accident or by diagnosis of, for example, Alzheimer's Disease, arthritis, depression, diabetes, heart disease, mental illness, multiple sclerosis, Parkinson's, and other conditions. No geographic location is immune from this prevalence, particularly in relation to the vicissitudes of contemporary life—for example, travel, trudgery, terrorism. Unsurprisingly, both poverty and place of residence both increase and exacerbate the vulnerability to being born with or acquiring a disability in one's lifetime. My focus on "Radical Bioethics" offers one response to a dearth of theo-ethical and bioethical reflection on a critical concern for this population. This concern includes the requirements of justice that have been largely ignored in the vein of care for the support and development of basic human functioning capabilities that are available in the common good, the means of which "commons" would be distributed with a preferential justice and safeguard for persons and communities of people with disability [1].

The ethics of critical medical and essential health care for people with disability remains under-considered explicitly, except perhaps as an aside to the focus of research in bioethics and healthcare ethics more broadly. Nevertheless, a growing number of academics in the humanities and social sciences recognize the need for bioethical, philosophical, and religio-theo-ethical reflection on the subject of disability. Many have begun to notice and reflect upon the ways in which people with disability have been ignored and/or grossly underestimated as participants in and rightful recipients of the commonweal: the common goods of health, education, recreation, employment, commerce, social and political affairs, and religious observance.

What follows continues work I have done on theological anthropology, exploring the diversity of creation, and the radical dependence that characterizes all people—not singularly or especially people with disability but—a dependence that characterizes all people, from the most robust to the infirmed. Such dependence raises critical questions of procedure in theological ethics, bioethics, and healthcare. Among the obstacles frequently encountered in this work are the persistent attitudes, notably identified by the social constructions of disability, that preclude many initiatives that would attend to both critical and basic health care for people with disability. Nevertheless, and even as most are not likely to have been exposed to the histories of the many tragic and rather appalling experiences that people with disability have endured, the good news is as important as the sad and scandalizingly brutal treatment people with disability have received at the hands of their caregivers, communities, and medical professionals in addition to abuse from strangers. Thus, as more people with disability participate today in many different settings where previously they had been excluded de rigueur, nondisabled people have become more accustomed to both casual and commercial interactions with them; my hope is to further the work of removing the obstacles to care and the inclusion of people with disability throughout the commons of human encounter.

In this essay I offer a broad look at the current lack of sustained attention in bioethics to the social forces and inherited assumptions that the field has failed to recognize particularly in an unexamined critique of its power and influence to shape the medical and social imaginary. Before entering into bioethics, I offer two precises, first a theological anthropology of radical dependence and second why radical bioethics –both viewed through a lens that includes an often-overlooked history of the experiences that people with disability have endured over the millennia. Theology undergirds the challenge of wholesale inclusion of people with disability in the main as well as in considerations of how people with disability are to be engaged in participatory bioethics like other moral agents in their exercise of autonomy and self-determination. I then explore the contours of a radical bioethics as it is rooted in principles with which I approach bioethics, namely, a radical posture and proclamation that people with disability are people first. I move then to difference and diversity as key to the root of the participation of people with disability in the commons and the professions. I conclude with a desiderata-like meditation on a radical embrace of flourishing for all.

2. A theological anthropology of radical dependence

People with disability and the non-disabled belong to a common humankind that is deigned the *imago Dei* and in which, through the *Kenosis*, God identifies with all human joys and sorrows, births and growth, disability and deaths, as well as the past and the future. As *imago Dei*, humankind takes its primary data from the Christian tradition's teaching on creation, the Incarnation, and Trinitarian theology. The implications of this theology extend beyond a facile nod to the identification of God with humankind: if the Trinity is the theo-anthropological ground of being human, then the Trinitarian symbol of God must function as a template for all divine and human affairs [2]. The critical importance of a Trinitarian theological anthropology cannot be overstated. If the datum of the tradition is to be believed, then we must be earnest in engaging with God and with one another along the lines of relationship and relationality that is the manifest expression of God for Us [3].

The Christian tradition has long discussed and defended the doctrine of the Trinity and the Trinity's action in the world as Father, Son, and Holy Spirit. However, the Trinity carries doctrinal importance beyond the formulaic introduction to prayer: the doctrine of the Trinity holds the potential of transformative practice among believers and nonbelievers alike. If every person is created in the image and likeness of God, and as God has been revealed to us in triune relationality *in se* and *ad extra*, then relationality—not isolating autonomy or dogged self-reliance—is the form of the *imago Dei*.

"Trinitarian theology is the language of relationality *par excellence*" [4]. not only *in se* but *ad extra* for us and for our salvation. The relationality that subsists within the Trinity is a relationality of a willed quasi-dependence of the Three Persons in One God. We Christians do not know God in any other way than in this Tri-Personal Relationality. This relationality is as God wills it to be in salvation history, *in se* and *ad extra* dependent in form and matter though uniquely unified and subsistent in effects: "the economic trinity is the immanent trinity and *vice versa*" [5]. It is thus *in se* wherein God enjoys/relates in God's own company and *ad extra* wherein God enjoys/relates to all that God has wrought for us—the symbol of a relational God functions.

"That symbols are signs is certain: they are expressions that communicate a meaning; this meaning is declared in an intention of signifying which has speech as its vehicle" [6]. Christian theology speaks of God in the terms of an I-Thou relationship. If the symbol of God functions in a vehicle of relationality and to be in relation suggests dependence within the one and the three, then relational dependence holds first place in the ways that we are to envision the *imago Dei*.

Of course, dependence in inter- and intra- Trinitarian relationality differs exponentially from the dependence experienced by the *imago Dei*, nevertheless dependence is in this anthropology relieved of its burdensome negative connotations. Dependence in this sense is neither a bad word nor a bad idea. It is, alternately, liberative. I am suggesting that unity in diversity translates to the *imago Dei* as radical dependence, human dependent relationality: we human beings are dependent upon God and upon one another from our conceptions and births, through childhood, adolescence and adulthood (to think otherwise is foolhardy) [7]. This symbol of God for us expresses intentional relationality as the ontological ground for understanding the function of the symbol that is the *imago Dei* rightly inclusive of every manifestation of God's creative incarnate sanctifying love or wrongly exclusive of any.

In this theology, relations exist on account of difference. God is conceptualized through a medium of difference or diversity as the Trinity of Divine Persons, different *and* related: God as Father and Creator, God as Son and Incarnate Word/Jesus of Nazareth/the Risen Christ of Faith, and God as Holy Spirit and Transcendent Love within the Godhead and to all the world. "The mystery of God is revealed in Christ and the Spirit as the mystery of love, the mystery of persons in communion who embrace death, sin, and all forms of alienation [e.g., difference] for the sake of life" [8]. In Christian terms, the only way for us to know God is through God's own self-disclosure as the divine being in relation with Godself in Trinitarian union and in relation with the rest of creation and perhaps we human beings in particular, in anthropological terms. As beings created in God's own image, we human beings are known only in relation to God and to one another: our self-knowledge is dependent on the diversity of persons and all other beings and things in the world.

I propose that we take difference or diversity as the key to our being in the world and in relationality with all reality thereby. While we may hesitate to consider disability as the sine qua non condition of humankind, it behooves us to recognize diversity and the dependence that attends to all beings and to all of our relationships, of intimate and of distant or impersonal kinds. This dependence on "the other"—all too frequently assumed true especially (only?) for people with disability—unfolds in both deliberate and indeliberate ways, by choice or by literal or figurative accident. However, whatever our present status on the ability- disability spectrum, our task is to become ever more mindful of the other(s) among us and to recognize the webs of connection in place of potential self-loathing or of rejection, disdain or fear of the unfamiliar/not related/unrelatable other. Such a potential is a blasphemy against God in whose image we are all created and by whose begetting us makes us members of a single family, more than neighbors, sisters and brothers all. And we need to confess the sins and the near occasions of these blasphemous sins against God, ourselves, and our siblings near and far. The radical dependence of being in the world belies autonomy and the "self-made man."

Thus, when we turn to disability identity with a posture of humility, we may soon discover the magnificent diversity in the ways that persons become themselves and we may soon find them and ourselves beautiful. Consider the tendency of delight many of us experience at the sight of a majestic mountain scape, a field of wildflowers, a herd of buffalo, a night sky filled with stars, or the songs of wild birds. We marvel at nature's diversity, but we may be stingy in recognizing diversity in humankind. In the world of dualistic segregations, superficially identifiable differences have been used to categorize and, invariably, establish hierarchies that ranked individuals and communities on the basis of their conformity to a norm. In the case of human norms the dualisms of male/female, spirit/body, white/non-white, heterosexual/homosexual, and non-disabled/disabled have designated de facto the second part of each pair as a defective version of the first part; those designations subsequently led to the oppression or patronization of the second by the first. However, when diversity, inclusive of people with disability, is presumed as normative, these dualisms lose their power to elevate one expression of diversity, however narrow or large, over the diversity of other expressions. When diversity is normative dualisms no longer make sense and an anthropology of inclusion can emerge in their place.

3. Why "radical" bioethics?

3.1 Meanings and implications of Radicality

The *Oxford English Dictionary* offers three meanings of the term "radical," each referring to the defining nature of someone or something as 1) fundamental or basic, essential, quintessential; 2) inherent or innate, intrinsic, structural; and 3) comprehensive or constitutive, organic, root [9]. I start with the notion of "radical" as root to agitate any personal complacence toward what we in the "west" have inherited as a hierarchy of being, at least since Linnaeus in the eighteenth century if not well before, with the 6th century BCE pre-Socratic philosophers to the Aristotelian trajectories in metaphysics of the 4th century BCE. Additionally, with the anti-racist, feminist, and LGBTQ critiques, I hold a view toward a socio-political and theological kind of disruption about the taxonomic hierarchy and the subsequently normative ways of thinking about ourselves as members of the family

DOI: http://dx.doi.org/10.5772/intechopen.97654

hominid, genus *homo*, and species *Homo sapiens*. I challenge determinations at the root of "Who Counts" as members of the human fold at the fundamental, inherent, and comprehensive levels of this hierarchy. And I challenge the agency that some members of the species have exercised in determining restrictively the agency of most others—those many others classified as marginal to the social, political, and religious, let alone the academic and professional elite.

I am suggesting a radical/root change in the way people with disability are perceived by many among the nondisabled community and the subsequent ways in which they—people with disability—are disabled by the social, medical, philosophical, political, and theological constructions of non-normative "being" in the world. The idea of constructions that cohere with the now widely accepted rejections of and efforts to dismantle racial, ethnic, and gender biases is critical to the work of dismantling stereotypical assumptions about disability and about people with disability. Briefly, the social construction approach presents the contemporary critique of long-held-to-be true determinations about individuals and groups of "like" individuals such that all persons belonging to the class have uniform experiences of "being" a Woman or Black or Indigenous or Gay ... or Disabled. Moreover, individuals and the groups to which they have been "assigned" are stigmatized for being ... women, non-Anglo, Native/Indigenous, Queer, and/or Disabled [10]. Thus, following the lead of people with disability and their co-agitators in the radical disability movement, my approach to disability has matured from a focus on individual problems experienced by individual persons with this or that particular impairment "to the wider oppression and social barriers that [have historically] excluded and disabled people" [11].

For all its efforts to promote autonomy, beneficence, nonmaleficence, and justice, bioethics is not immune to the inherited social constructions of disability. These constructions hold alongside the inherited assumptions about people with disability and the various ways in which they, people with disability, have been and continue to be calcified, categorized, and classified as, mostly, unfit to share space with the nondisabled. However, since none of us are immune from these inherited assumptions and their subsequent applications to real people, it is important to unpack the assumptions about disability for the dangers that lurk within them. Similarly, it is important to dismantle the oppressions that accompany the constructions that build on these assumptions. Granted, none of us likes to think that we can be mistaken about the values inherent to this or other concepts (like beauty, strength, and adaptability). Nevertheless, glaring examples of the misappropriation of personhood abound: for example, some of the handing down of religious practices from generation to generation were conducted under penalty of death if refused and, just as mistakenly, imperial conquests (e.g., by the Portuguese, Spanish, British, French, Chinese, Ottomans, and the United States) denied the humanity of many indigenous peoples in Africa, the Americas, and Asia. Let us not be fooled. Many people today suffer enslavement in the form of human trafficking for exploited labor, organs, or reproductive service, and sex [12]. In ways similar to the contemporary enslavement of people who occupy places hidden from a decidedly prejudicial social history and from "polite" company, too few people register a thought about the similarly prejudicial, marginalizing, and oppressive experiences of people with disability.

With exceptions, people with disability have not been treated well. Their treatment has been identified and outlined for us by social science academics, psychology and nursing professionals, and humanities scholars in a system of models that distinguish one manner of treatment–with positive or negative effect—from another manner. Contemporary studies offer an approach to this history through the models of disability, developed by people with disability, that are related closely to the ways in which the non-disabled and dominant codified their perceptions of people with disability according to the social roles to which they were assigned. The most common models of disability are the religious-moral model where individuals or their parents or communities are responsible for disability as a punishment from God for sin or sins committed; the medical model, which "conceptualizes disability as deviance and lack within the individual, and therefore all medical interventions are geared toward bringing the individual as close to normalcy as possible" [13]; and the social construction model of physical and attitudinal barriers of exclusion, like stairs and inaccessible educational, commercial, health, political, or recreational opportunities. In addition to these three, other precisions have been offered with the tragedy/charity, expert/professional, rehabilitation, economic, and rights based, customer/empowering models [14]. Parallel to these models are social roles to which people with disability have been assigned as their being sick or sub-human, a menace, pitiable, a burden, holy innocent, inspirational, amusing, and a blessing [15]. Regardless of model or role, each of these assignments includes greater or lesser degrees of stigma: the defining mark of otherness that clears the way to marginalization and to greater or lesser degrees of direct marginalization, oppression, and violence [16]. The models offer a shorthand reference to understanding the presumptive attitudinal barriers that people with disability and their companions encounter all-too-frequently to this day [17].

As disability advocates remind us: "it's important to remember here that throughout recorded history all forms of inequality, injustice, and oppression have been sanctioned in one way or another on the basis of assumptions of biological inferiority" [18]. Contemporary efforts to decry these injustices and to reject these assumptions, such as to include the perspectives and insights of people with disability, are rare and, when present devolve all-too-easily into patronizing thanks and nods. "Why, at almost the end of the second decade of the 21st century, are the [basic and fundamental] human rights of people with disability still ignored?" [19].

As suggested above, those with the power to make and shape societies have been grossly mistaken in their judgment about the inherent human value and dignity belonging to people with disability and others who do not conform to hegemonic norms. Those mistaken judgments are the bases of a history of maltreatment that people with disability have endured, a history that has been largely ignored and likely intentionally unrecorded; in effect, people with disability themselves and their stories of success and failure and of loves and losses have been silenced over the course of time. However, that culture of silence is no longer acceptable: the truth to be told is that newborns, infants, children, and adults have been neglected, abused, and exterminated on account of the presence of disability in their lives [20].

I now turn to the historically prevalent exclusion of people with disability to contextualize their experience and its residue in the main and in bioethics today.

3.2 A brief survey of historical experiences

With 15% to potentially 25% of people worldwide having one or more disability today (up to 1.75 billion of 7.6 billion people), it is undeniable that people with disability have been among the members of the human economy from antiquity to the present. Combining the models of disability (medical, moral, and social) to parallel social roles (menace, burden, clown), individuals were identified taxonomically as other. Given the lessons that contemporary retrievals of the historical experiences of many members of minority populations have uncovered, this "othering" of people with disability has resulted in their oppression as a class that, like racism and sexism, can no longer be tolerated.

DOI: http://dx.doi.org/10.5772/intechopen.97654

To the extent that those who have held power and authority record history, resolutions concerning people with disability resulted in their marginalization on account of the causes those powers presumed were at fault from: divine punishment for some sin (either one's own or one's parents) or a pre-emptive warning show of divine power; consorting with evil; an imbalance of humors; maternal stress during pregnancy; bestiality; menstruation; and astrology. Each of these causes encouraged perceptions that people with disability were more animal or otherworldly than human, that they could tolerate environmental extremes and malnutrition, and that they were dangerous to the societies in which they lived—conclusions that gave license to harm them with impunity by taming, exhibiting, sequestering, and worse.

Scandalously, many individuals with disabilities–feared and/or loathed by the non-disabled—would have been exposed at or near birth or otherwise ostracized once the presence of a disability became known [21]. Infanticide by exposure was widespread and, in some state-sponsored cultic systems, the practice was mandatory. Some early Greek medical texts instruct on recognizing defects at birth, in the first months, and early years so as to determine a child that is not worth raising [22]. Aristotle too recommended laws to prevent the rearing of "deformed" children and to deny deaf children access to schools, since they would burden the progress of the non-disabled children/boys of the community [23]. In Greco-Roman antiquity it would not have been uncommon for newborn girls or a newborn with observable disability to be abandoned or left in a crude cradle at a crossroads or near a market, gymnasium, or temple with some possibility of being taken (and likely enslaved), or tossed into a river by their patriarchs. Equally troubling and perhaps more horrifying, some parents or overseers and other wardens, who depended on "income" from begging, would mutilate their biological or "adopted" children with disability to increase the pity-value that patrons might assess on them and thereby increase almsgiving to their cause of household maintenance [24]. Among other curiosities, the Roman gladiator games included the spectacle of fights between little people, the deaf, other people with varying disability, women, and animals. Less brutal but not less disturbing were practices that exhibited individuals with disability in courts of power: as a sign of blessing, entertainment, or pity to extend telethon-like charity.

The early medieval period made way for the custom of caring for the sick, those with disability, and the poor. Outside of the support of their natal homes, people with disability were reduced often to poverty and they resorted to begging as a principal means of income. Wanting to follow the example of Jesus, who attended to those who were marginalized for this or that stigma, Christians began to extend compassion on the less fortunate. By the height of the Middle Ages in Europe, a "period of organized beggary" led to guilds open to people with disability wherein leaders emerged, and rules and languages developed by guild members [25]. The guilds represent a welcome initiative by today's standards. Yet, this same period saw the institution of "idiot cages" that kept people with disability confined, while the cage protected those beyond the bars. And where cages were insufficient or when the masses tired of this or that caged group, the "ship of fools" provided another form of distance to keep people with disability separate and exploited as members of a traveling carnival-horror-freak sideshow for port residents and visitors alike.

And then came the development of institutions. Founded as a result of a system of hostels for pilgrims on their way to a holy site for both blessings and cures, hospitals for the sick and incurable became asylums for the insane and invalid. With the advent of the Enlightenment project to reject the old and quaint in favor of a rational order, new scientific ways of conceiving the individual in society and the common good brought to the fore utopian concerns of a more perfect communion, overtly including an underlying concern for the dangers lurking in any near presence of people with disability. The isolation of institutions provided safety for the non-disabled as well as it gave rise to better or worse care for those institutionalized where—up to and including the 21st century—as a captive population, they could be studied "objectively." With concentrated access to people with a diverse array of disabilities, doctors and scientists began to investigate the causes of disabilities using then newly advanced medical and empirical methodologies. Some of this early science fueled the later nineteenth and early to mid-twentieth centuries' eugenics movements through the subsequent sterilization of people with disability and other suspicious folk [26]. Consider the scientific "proofs" of a biological basis for the categories of race and the subsequent discrimination against non-white peoples, especially peoples of African descent, that labeled many deviant. People from Mediterranean countries and Asia were considered to be of "questionable genetic stock" and likely to increase the number of feebleminded or criminals that would become wards of the state; it would be better to prevent them from reproducing altogether [27]. As long as people were institutionalized and isolated from general human commerce, they were–and those who remain institutionalized are—vulnerable to abuse, exploitation, and other dehumanizing injustices.

The confluence of social progress, science, and rational self-interest led to the systematic individualization and medicalization of all persons–those deemed "normal" and those deemed "othered"—as subjects (the normalized) and as objects (the aberrant/abnormal/disabled). This systematic program led to a widespread ideology of disdain for, dis-ease with, and distrust of any who did not/do not conform to the hegemonically putative/normative/ideal modern man. By the twentieth century, eugenic initiatives were set in Europe and the United States with sterilization programs and final solutions in a murderous holocaust of untold, unnumbered, and unaccounted hundreds of thousands of people with disability. Scandalously still, eugenics and euthanasia by a different name continue apace with neonates, children, adolescents, and adults in their prime and elderly with disability as today's principal populations that are vulnerable to medical-social-scientific control. While eugenics may not be institutionalized, it holds ideological power and is practiced widely in reproductive medicine and the selective abortion of fetuses. Similarly, euthanasia remains a threat as the contemporary equivalent of exposure by withholding life support from a person–neonate or adult—who could thrive if given the chance, not with heroic or extraordinary intervention but with the radical bioethics notion of ordinary care.

4. A disability-informed bioethics

4.1 Disability is multi-dimensional and multi-experiential

"Disability is a multi-dimensional concept, which should be understood in terms of a continuum" [28]. This continuum is true for all people once born and throughout the days of our lives. Even so, a disability continuum may have more dramatic punctuations than the general population. As such, bioethics attention to the multifaceted experiences of people with disability has the potential to integrate disability experiences in both critical interventionist care for things like substance abuse or cancers as well as the more mundane and presumably easier access to routine health checks and preventive holistic services like nutritional support, exercise, education, and social interaction in arts and leisure and recreation. Bioethics will need to approach the subject with humility, since any attempts to categorize disability in generic terms will fail, especially since the human organism is itself complex; nevertheless, the phenomena of disability are expressed in the literature as physical impairments, sensory impairments (e.g., blind and deaf), cognitive and/or developmental difficulties, mental health, and chronic illness.

While disability has been a feature of human life throughout the millennia, the contemporary climate suggests that the phenomena is rare or, if not rare, better to be left unspoken and closeted. The history above belies the rarity of disability and the suggestion that a culture of glamor or power or the accumulation of wealth is sufficient to disguise the presence, the challenges, the joys, the hopes and sorrows, as well as the contributions of persons with disability in ways grand and small yesterday and today [29]. The initiatives of the United Nations in its *Convention on the Rights of Persons with Disabilities* (2006) are key for advancing the cause of recognition and self-determination for all people with disability. Although the United States has signed, it has not ratified the *Convention* and remains thereby not bound by its statutes. Article 4 notes nonetheless that "States Parties undertake to ensure and promote the full realization of all human rights and fundamental freedoms for all persons with disabilities without discrimination of any kind on the basis of disability" [30]. The full realization of all human rights requires that persons with disabilities, like the nondisabled, have access to the basic goods of safe housing, potable water, nutritious food, education, family relations and friendships, healthcare, employment, recreation, public services, and religious or other spiritual practice. Additionally, Article 10 reaffirms "that every human being has the inherent right to life and [States Parties] shall take all necessary measures to ensure its effective enjoyment by persons with disabilities on an equal basis with others" [31].

In 2001 the World Health Organization published the *International Classification of Functioning, Disability, and Health* (ICF) and in 2007 the *ICF-CY* (children and youth) as the framework for measuring health and disability at both individual and population levels. In these texts and in the related 2011 *World Report on Disability*, WHO conceptualizes a person's level of functioning as a dynamic interaction between her or his health alongside environmental and personal factors, with a comprehensive basis for the definition and measurement of health and disability [32]. The idea of functioning as a measurement standard was inspired by the work of economist Amartya Sen and philosopher Martha Nussbaum, who developed the "capabilities approach" to discern an individual's functional development and attainment of health [33]. In brief, the capabilities approach holds that all human beings have a virtual obligation to develop the abilities–inherent to each albeit in variable and disproportionate measure across the capability spectrum— as a positive natural right and the province of human initiative that lead, if given the opportunity, to flourishing and a good human life. Nussbaum articulates these capabilities as the basic human rights to acquire functional development of life: bodily health; bodily integrity; senses, imagination, and thought; emotions; practical reason; affiliation; concern for other species; play; and control over one's environment [34].

As used in the ICF, the capabilities approach offers a holistic metric to determinations of health and well-being based on individuals' development of abilities aligned with the personal, local, regional, national, and global infrastructures—of educational, occupational, medical, recreational, and social opportunities—that are necessary to support that development. In matter-of-fact straightforwardness, WHO admits "Disability is part of the human condition" [35]. WHO argues further:

> *[The] ICF is named as it is because of its stress is on health and functioning, rather than on disability. Previously, disability began where health ended; once you were disabled, you were in a separate category. We want to get away from this kind of thinking. We want to make [the] ICF a tool for measuring functioning in society, no*

> *matter what the reason for one's impairments. So, it becomes a much more versatile tool with a much broader area of use than a traditional classification of health and disability.*
>
> *This is a radical shift. From emphasizing people's disabilities, we now focus on their level of [functioning and] health [36].*

4.2 A radical shift in understanding disability

The ICF distinguishes between body functions, body structures, activities and participation, and environmental supports or lack thereof. To use the language of more common parlance: the functions reflect the purpose of mental, sensory, voice, organ, metabolic, reproductive, neural-muscular-skeletal, and skin systems; the structures refer to the engagement of procedures or steps involved with voluntary and involuntary movement. Activities and participation consider the degree to which individuals engage both functions and structures from cognition, affect, and locomotion, to self–/family–/community–/social–/civic-care. Environmental factors include considerations of the presence or absence of support for integral human development and flourishing.

Thus, given the complexities of functions and structures, disabilities fall into one or multiple classifications. In a similar vein, many people have co-occurring symptomatic dysfunctions, particular disabilities, and health complications with their primary mental, sensory, voice, organ, metabolic, muscular-skeletal disability. Under the Americans with Disabilities Act (ADA), the categories that qualify a person for accommodations of individualized support or relief are expressed in physical or mental impairments that interfere with major life activities [37]: Affective Disorders, Autism, Blindness, Cognitive Disability, Deafness, Emotional/Development Delay, Hearing Impairment, Intellectual Disability, Muscular/Physical/Skeletal Impairment, Neurological Impairment, Other Health Impairment, and Specific Learning Disability. These initiatives and legal precedents cohere with a baseline understanding of human capability that takes the contexts and particularities in which individuals and communities live as key to unlocking and supporting everyone's basic human functioning capabilities. Nevertheless, "People with disability are characterized by low human and social capital" [38]. Thus, to consider health on the basis of functional capabilities development is both promising and dangerous for people with disability. Promising since focus is placed on the determinations of an individual's capabilities and efforts in collaboration with social systems to develop those capabilities; dangerous since location will determine access to those necessary support systems. As a cause and consequence of disability, poverty remains the single most difficult obstacle to overcome, and poverty is directly related to an individual's ability to both develop and then exercise her/his basic functioning capabilities and thereby to thrive.

5. The difference of radical bioethics

The thing I propose as "radical" here is not in the sense of "protocol-be-damned" but in the more mundane and more nuanced frame of the ways in which a fundamental set of attitudes and actions can take hold in matters pertaining to bioethics in general, to the subject of disability and, more importantly, toward persons with disability in particular. The radical nature of this inquiry hearkens to the origins of the discipline of bioethics begun with the Hastings Center in New York (1969),

DOI: http://dx.doi.org/10.5772/intechopen.97654

the Kennedy Institute for Ethics at Georgetown University (1971), and to Van Rensselaer Potter (1970), the oncologist who coined the term (at least in its English usage) [39]. Potter was particularly interested in the intersections and shared information of findings between the biological sciences and the humanities so as to ensure the benefits of research would yield results that attend to real persons and the eco-systems that support life: "global in scope, transdisciplinary in method, and, most importantly, compelled by a commitment to action that demanded personal engagement with social issues" [40]. This inquiry is radical in its adherence to the foundations of the discipline per Potter and our early colleagues at the Hastings Center and Kennedy Institute—many if not most of whom were trained as undergraduate students in philosophical and/or theological disciplines and for whom the sciences of medical care and interventions were perforce designed for human health and the social good. Further, my project invites you to adopt this interdisciplinary approach of a radical bioethics of dependence on the whole sphere of human commerce, with dependence as a normative key thereby for all persons inclusive of persons with disability across the millennia.

Difference is key to appreciating the diversity of persons and the perhaps even greater diversity of experiences among people with disability as equal to those among the general population. In order to ensure a comprehensive view, the insights of sociological critiques, which approach bioethics with quasi casuistry from the particularity and context of specific cases, offer a compelling argument that attends to lived experience, institutional culture, and structural injustice as the starting places to uncover the realities that honor persons with disability [41]. These approaches recognize that determinations of functioning capabilities depend upon considerations of interpersonal relations, institutional structures, and the overall social world wherein the subjects of concern, whether persons with disability or women or people of color, are situated in real time and place.

Moreover, "We cannot reduce the complexity of disability to either a biological problem, a psychological problem, or a social problem" [42]. All the factors of an individual's life must be considered and interventions—of medical and rehabilitation kinds, assistive devices, psychological support, barrier removal, welfare benefits, legal protections, and cultural change—must be engaged at different and particular levels for the benefit of the individual in need of care (there is no one protocol that fits all persons adequately). As many in the field of Disability Studies argue forcefully, disability may present as a health concern, but it is more an issue of social and economic concern. As noted above, across the world people with disability lack access to basic health and rehabilitative services as well as a lack of social support in the development of their basic human functioning capabilities. "They face barriers and prejudice, or poorer quality of healthcare. This [subpar access to care] means their health outcomes are worse—not as a result of their underlying impairments, but because of failures of [access to] general care" [43].

But what is difference at its root? Difference is a condition of being or a relation of distinction or diversity between one thing or person and another. An older Latin connotation points difference in the direction of diversity and is suggestive of variety, a point of dissimilarity but similar enough as to be recognizable as this or that thing or person. I have argued that diversity is the distinguishing feature of all creation, human beings included [44]. And I argue that diversity is the signature of God's handiwork throughout the known world and beyond. I recognize this diversity as God's own "calling card" and the way that God, in the Christian tradition, reveals Godself to us in relation with self and with others.

6. Radical bioethics: desiderata

I have long avoided the question of desire in Ref. to interrogating my own life and its wondrously circuitous and amazing turns except to ponder the opportunities given alongside the choices made that brought me to this moment in time, and to give thanks. I've led a charmed life, not without roadblocks here and there but, charmed, nonetheless. For an even more conscious period of time, I have avoided the question of desire in Ref. to the lives of people with disability. I like to think that this avoidance is rooted in a posture of humility by virtue of not knowing in any intimate sense what life is like in another person's shoes. I cannot truly fathom another person's "longing for something lost or missed," "to feel the loss of," or "to be wanting" [45], the root meaning of desire. Nevertheless, I am drawn now into this subject in recognition of the sad history of medical and social treatment that people with disability have experienced across the millennia—from exposure to bullying, abuse, and murder—and to conversations that many academics, medical professionals, and policy makers have regarding the spoken and unspoken assumptions that "they"/people with disability would be better off dead. I am drawn also into the subject of desiderata by genuine calls initiated by some persons with disability and their family members, friends, and caregivers for interventions that promise rightly to relieve some of the conditions—especially physical pain and the internalized suffering of rejection—that compromise human flourishing.

"Disability, in everyday thought, is associated with failure, with dependency, and with not being able to do things. [Many of the nondisabled] imagine it must be miserable to be disabled. [But] both empirical evidence and anecdotal testimony reveals that for many people with disabilities, life is surprisingly good" [46]. Moreover, when asked, many people with lifelong disability say, "we don't want to be cured." For many, except for bouts of pain (it is initially very different for people who acquire disability in their teen and adult years), their lives are fulfilling the way they are and their disability is part of their identities such that it is near impossible to envision a life without disability. "The medical focus of cure and change [is] linked to an assumption that disabled people want to be cured. ... The dominant discourse fail[s] to recognize disability as a desired differentness, which can be core to an individual's identity" [47].

Despite the best of intentions of family members and caregivers, like the nondisabled, persons with disability are themselves the principal subjects of their own lives and desires and they are thereby entitled to the exercise of autonomy. I am presuming a degree of cognitive and communicative autonomy that may be absent on account of age or developmental disability. Age aside, unfortunately, their desires were rarely taken into account across the standard practices of paternalism, such that past discussions in the medical arena on life with disability were often limited to questions of "to treat or not to treat" (and to let die) [48]. Today, with increasing Disability Rights Advocacy and given the voices of people with disability on the subject of cure, questions of intervention point more directly toward facilitating life with disability through barrier removal alongside of relief for sickness when autoimmune responses or influenza or cancer or diabetes or other calamity present. Thus, not unlike preferred choices when it comes to the dinner menu, decisions regarding this or that intervention, care protocol, or cure demand the exercise of personal autonomy and must be solicited from persons with disability in as equally an informative measure and accessible language as are decisions solicited from the nondisabled: the desires of people with disability must be honored.

As I return then to desiderata, a minimum desire among the communities of people with disability is to recognize their agency. Granted, the spectrum of conditions that qualify as disabling are themselves diverse and often overlapping, both identity and agency diversities will emerge between physical, cognitive, and developmental disabilities but social stigma—the historically definitive construction of people with disability as inferior "to the main" and, as such, "other"—remains a common experience across the spectra. The de-construction of stigmatizing "otherness" remains the principal desiderata of my work. From that recognition, the tangible desires voiced by people with disability include minimally [49]: 1) the removal of barriers both physical and attitudinal to wholesale inclusion throughout the many avenues of social commerce—oppression remains the single most problematic of personal barriers to overcome; 2) reasonable accommodations to facilitate participation (e.g., accessible communication formats like sign language, braille, and illustrations); 3) an overall slower pace in language and in movement from place to place and for task to task; 4) attention and equitable access to basic and critical healthcare, education, and employment; and perhaps most of all 5) friendship and other personal relationships beyond kith and kin in educational, social, commercial, employment, political, recreational, and religious arenas.

What else is to be desired?

Since many disabilities are acquired over the course of a person's life, prevention of disabling impairments is an obvious desire. Hence, practical initiatives to reduce acquired disability include hard hats and other protective gear for manual laborers, immunizations from communicable diseases, reduction of exposure to hazardous materials, balanced diet for all and nutritional support especially for women who are pregnant or planning pregnancy, moderate alcohol and other "recreational" drug consumption, respect for speed limits and traffic conditions, avoidance of violent games and guns. We delude ourselves into thinking that any of us are immune from any of these eventualities.

On immediately practical and tangible levels, people with disability globally face obstacles to living and to living well that many of us see only as voyeurs during newscasts or from charity appeals for help. But the daily needs of most people with disability are embarrassingly simple: mobility aids like wheelchairs and rollators, barrier removal, reliable electricity, access to potable water, protein and carbohydrate, contained human and animal waste, basic healthcare and primary education. Let us recognize that meeting the desires of people with disability is not rocket science but, until we break free of these attitudinal and ideological barriers to recognition that people with disability are people first, these basic needs of common goods remain out of reach. It's rather simple to admit, but not easy to make the changes necessary for a tomorrow that all can better enjoy.

7. Concluding thoughts on radical bioethics

I started this work with an invitation to consider today's more than one billion people with disability as one of the most diverse populations—15% to potentially 25% of the 7.6 billion people worldwide. This population is a mass of people relegated to the margins of the larger social groupings to which they belong. In a time when gender, race, and bio-ecological diversity are championed and barriers to inclusion dismantled for some, the margins of human commerce to which many if not most people with disability are consigned are no longer tolerable.

Truth be told, medical and healthcare professionals have approached care concerning persons with disability with a jaded view and a jaded past regarding their worth as marginal at best or their status as less than deserving of either routine or critical care. In response, I suggest, this "radical bioethics" that 1) invites healthcare and bioethics professionals to recognize that a patient with disability is a patient first; 2) points to a lack of attention on the part of these professionals and the discipline of bioethics to be aware of the similarities and differences that disability presents in deliberations of treatment protocols; and 3) as WHO admits, "from emphasizing people's disabilities, we [have made a radical shift to] now focus on their level of [functioning and] health." Thus, following the lead of members of the communities of disability who have engaged legal argumentation on behalf of their vulnerable sisters and brothers with disability who have not received a fair hearing regarding their care, this disability consciousness is best informed before considerations about medical interventions available to persons with disability are pronounced.

As Professor of Law and Bioethics Alicia Ouellette observes:

"bioethicists tend to support individual choices to refuse medical care, family decision making, and advance directives. Members of the disability community are often skeptical of or opposed to these practices. Some disability experts view medically assisted nutrition and hydration as a basic human right; bioethicists tend to think of medically assisted nutrition and hydration as no different from other medical treatments [that may be withheld or withdrawn]. Bioethicists support efforts of doctors to "fix" physical impairments; disability scholars question the need to "fix" the bodies of individuals with disability and look instead for societal solutions. Many bioethicists view persistent vegetative state as something entirely different from other disabling conditions; some disability activists deny those differences. [And] many people in bioethics seek to resolve individual cases without taking into account social and community concerns, whereas social and community concerns are central to the disability community" [50].

Desiderata.

And a final word regarding the title of this work. Philosopher and lawyer Max Ehrmann wrote the poem "Desiderata" in 1927, it was published posthumously in 1948. Its popularity may have waned of late but we human beings continue to burn with desires/*desiderata* of many kinds, some mundane and others profound. Whatever the desires of people with disability, their family members, friends, those who care for and about them, and those who do not, I think we can take Ehrmann's word to head and heart in our strivings for a better tomorrow:

> *Go placidly amid the noise and haste, and remember what peace there may be in silence.*
> *As far as possible, without surrender, be on good terms with all persons.*
> *Speak your truth quietly and clearly;*
> *And listen to others, even to the dull and ignorant; they too have their story.*
> …
>
> *Beyond a wholesome discipline, be gentle with yourself.*
> *You are a child of the universe, no less than the trees and the stars;*
> *[And] you have a right to be here.*
> …

DOI: http://dx.doi.org/10.5772/intechopen.97654

Author details

Mary Jo Iozzio
School of Theology and Ministry, Boston College, Boston, MA, USA

*Address all correspondence to: mary.jo.iozzio@bc.edu

References

[1] This work was sponsored by the Loyola Marymount University Bioethics Institute and is a revision of the Annual Lecture of the Austin and Ann O'Malley Distinguished Faculty Chair in Bioethics. Los Angeles, CA, 2019

[2] See Johnson E. She Who Is: The Mystery of God in Feminist Theological Discourse. Crossroad Publishing Co, New York; 1992

[3] See LaCugna, CM. God for US: The Trinity and Christian Life. Harper Collins, New York; 1991

[4] LaCugna CM. Reconceiving the Trinity. Scottish Journal of Theology. 1985;**38**(1):13

[5] Rahner K. The Trinity. Herder & Herder, New York; 1970. p. 22

[6] Ricouer P. The Symbolism of Evil. Beacon Press, Boston; 1969. p. 14

[7] See Iozzio MJ. Radical Dependence and the Imago Dei: Bioethical Implications of Access to Healthcare for People with Disabilities. Christian Bioethics. 2017;**23**(3):234-260

[8] LaCugna CM. God For Us. 1

[9] Oxford English Dictionary/OED Online, 3rd ed. Oxford University Press, Oxford; 2019. s.v. "radical"

[10] See Goffman E. Stigma: Notes on the Management of Spoiled Identity. Prentice Hall, Upper Saddle River, NJ; 1963

[11] Shakespeare T. Disability: The Basics. Routledge, London; 2018. p. 13

[12] See Zimmerman C and Kiss L. Human Trafficking and Exploitation: A Global Health Concern. PLoS Med 2017;**14**(11): e1002437, https://journals.plos.org/plosmedicine/article/file?id=10.1371/journal.pmed.1002437&type=printable; Zimmerman C, Hossain M, and Watts C. Human Trafficking and Health: A Conceptual Model to Inform Policy, Intervention, and Research. Social Science & Medicine 2011;**73**(2): 327-335; Campbell LM, Zimmerman YC. Christian Ethics and Human Trafficking Activism: Progressive Christianity and Social Critique. Journal of the Society of Christian Ethics. 2014;**34**(1): 145-172

[13] Erevelles N. Disability and Difference in Global Contexts: Enabling a Transformative Body Politic. Palgrave Macmillan, New York; 2011. p. 19

[14] See Shakespeare T. Cultural Representation of Disabled People: Dustbins for Disavowal? Disability & Society 1994;**9**(3): 283-299; Marks D. Models of Disability. Disability and Rehabilitation 1997;**19**(3): 85-91; Llewellyn A and Hogan K. The Use and Abuse of Models of Disability. Disability & Society 2000;**15**(1): 157-165; Masala C and Donatella Rita Petretto DR. Models of Disability, in Stone JH and Blouin M, eds. International Encyclopedia of Rehabilitation (2010), http://cirrie.buffalo.edu/encyclopedia/en/article/135/; and Oliver M. The Social Model of Disability: Thirty Years On. Disability & Society. 2013;**28**(7): 1024-1026

[15] See Wolfensberger W. A Brief Overview of Social Role Valorization: A High-order Concept for Addressing the Plight of Societally Devalued People and for Structuring Human Services. Valor, Plantagenet, Ontario; 2013 [The Training Institute for Human Service Planning, Syracuse, NY 2004]

[16] See Goffman E. Stigma: Notes on the Management of Spoiled Identity

[17] See Masala and Petretto. Models of Disability

[18] Barnes C. Disability Studies: What's the Point? Presentation notes. Disability Studies: Theory, Policy, and Practice. September 4, 2003. University of Lancaster. p. 20

[19] Jackson MA. Models of Disability and Human Rights: Informing the Improvement of Build Environment Accessibility for People with Disability at Neighborhood Scale? Laws. 2018;7(1): 2

[20] On the ramifications of a culture of silence, see Freire P. The Adult Literacy Process as Cultural Action for Freedom, and Cultural Action and Conscientization. Harvard Educational Review. 1998;**68**(4)

[21] See French RS. From Homer to Helen Keller: A Social and Educational Study of the Blind. American Foundation for the Blind, New York; 1932

[22] Garland R. The Eye of the Beholder: deformity and disability in the Graeco-Roman world. Barnes & Noble Lyndhurst, NJ/Bristol Classical Paperbacks; [1995], 2010

[23] See Aristotle, Sense and Sensibilia 1 (437a16). Beare JI trans. Classics MIT, Boston; The Internet Classics Archive, Daniel C. Stevenson, Web Atomics, 1994-2009, http://classics.mit.edu/Aristotle/sense.html; Aristotle, Politics, Book VII, 16 (1335b20). Jowett B trans. Classics MIT, Boston; The Internet Classics Archive, Daniel C. Stevenson, Web Atomics; 1994-2009 http://classics.mit.edu/Aristotle/politics.html; and Seneca the Elder, Controversiae, IV, especially Mendici Debilitati. Winderbottomtrans. Harvard University Press, Cambridge; 1974

[24] Among others, see Parry M. From Monsters to Patients: A History of Disability. Arizona State University, Dissertation; 2013; and Boswell J. The Kindness of Strangers: The Abandonment of Children in Western Europe from Late Antiquity to the Renaissance. Pantheon Books, New York; 1988

[25] See Farmer S. The Beggar's Body: Intersections of Gender and Social Status. and Peyroux C. The Leper's Kiss, in Farmer S and Rosenwein BH, eds. Monks & Nuns, Saints & Outcasts. Cornell University Press, Ithaca, NY; 2000. 153-171 and 172-188

[26] See the discussion of "positive eugenics" in Jonsen A. The Birth of Bioethics. Oxford University Press, New York, 1998; 169-172; on the biological basis of racial discrimination see, for example, Washington HA. Medical Apartheid: The Dark History of Medical Experimentation on Black Americans from Colonial Times to the Present. Doubleday, New York; 2006

[27] See Cohen A. Imbeciles: The Supreme Court, American Eugenics, and the Sterilization of Carrie Buck. Penguin Books, New York; 2016

[28] Shakespeare. Disability, 5

[29] Among others, see Swinton J. Becoming Friends of Time: Disability, Timefullness, and Gentle Discipleship. Baylor University Press, Waco, TX; 2017; Burch S and Rembis M. Disability Histories. University of Illinois Press, Urbana; 2014; Erevelles, Disability and Difference in Global Contexts; Reynolds TE. Vulnerable Communion: A Theology of Disability and Hospitality. Brazos Press, Grand Rapids, MI; 2008; and Mitchell DT and Snyder SL. Narrative Prosthesis: Disability and the Dependencies of Discourse. University of Michigan Press, Ann Arbor; 2000

[30] United Nations, Convention on the Rights of Persons with Disabilities. New York, 2006; Art. 4.1; https://www.un.org/development/desa/disabilities/convention-on-the-rights-of-persons-

with-disabilities/article-4-general-obligations.html

[31] UN, CRPD, Art 10; https://www.un.org/development/desa/disabilities/convention-on-the-rights-of-persons-with-disabilities/article-10-right-to-life.html

[32] World Health Organization. World Report on Disability. WHO and The World Bank, Geneva; 2011; https://www.who.int/disabilities/world_report/2011/en/

[33] See Sen A. Collective Choice and Social Welfare: An Expanded Edition. Harvard University Press, Cambridge; 2017; Nussbaum M. Frontiers of Justice: Disability, Nationality, and Species Membership. Harvard University Press, Cambridge; 2006; and Sen A. Capability and Well-being, in Nussbaum and Sen, eds, The Quality of Life. Clarendon Press, Oxford; 1993; 30-53

[34] See Nussbaum M. Capabilities and Human Rights. Fordham Law Review. 1997;**66**(20):273-300 at 287-288

[35] WHO. First World Report on Disability. Geneva: WHO, 2011; 4

[36] WHO. Towards a Common Language for Functioning, Disability and Health: ICF. World Health Organization, Geneva; 2002; 3 https://www.who.int/classifications/icf/icfbeginnersguide.pdf?ua=1.

[37] I am here combining the terms of Federal Regulations regarding the Americans with Disabilities Act and the specific disabilities covered for children under the terms of the Individuals with Disabilities Education Act (re-authorized 2004, amended in Every Student Succeeds Act of 2015). See ADA. Law/Regulations. United States Department of Justice, Civil Rights Division; 2010, https://www.ada.gov/2010_regs.htm, and Individuals with Disabilities Education Act, "About IDEA," US Department of Education n.d. https://sites.ed.gov/idea/about-idea/

[38] Palmer M. Disability and Poverty: A Conceptual Review. Journal of Disability and Public Policy 2011;**21**.4: 210-218. On the agenda of capitalism and the impoverishment of people with disability and other minorities, see Erevelles, Disability and Difference in Global Contexts

[39] See Rensselaer Potter V. Bioethics: The Science of Survival. Perspectives in Biology and Medicine. 1970;**14**(1): 127-153

[40] Johnson J. Bioethics as a Way of Life: The Radical Bioethics of Van Rensselaer Potter. Literature and Medicine. 2016;**34**(1): 7-24 at 8

[41] Among others, see Sherry M. Disability and Diversity: A Sociological Perspective. Nova Science Publishers, New York; 2008; Cheng RP. Sociological Theories of Disability, Gender, and Sexuality: A Review of the Literature. Journal of Human Behavior in the Social Environment 2009;**19**(1): 112-122; and Erdman JN. Commentary: Bioethics, Human Rights, and Childbirth. Health and Human Rights Journal 2015;**17**(1) https://www.hhrjournal.org/2015/06/commentary-bioethics-human-rights-and-childbirth/

[42] Shakespeare, Disability, 21.

[43] Shakespeare, Disability, 88.

[44] See Iozzio. Radical Dependence and the Imago Dei. Christian Bioethics. 2017;**23**(3): 234-260

[45] See OED Online. s.v. desire, n., 1 and 5; s.v. desire, v. 4

[46] Shakespeare, Disability, 46

[47] Among others, see Beauchamp-Pryor K. Impairment, Cure,

and Identity: 'Where do I fit in?' Disability & Society 2011;**26**(1): 5-17 at 6; many online forums include discussions concerning the pursuit of cure or acceptance of life with disability. Among others, see eparent.com. Curing or Living with Disability: Your Choice Would be ...; Special Needs Resource. http://www.eparent.com/features-3/curing-living-disability-choice-commentaries/. Klenke M. How Society's Focus on 'Curing' Disability Can Harm Disabled People. The Mighty https://themighty.com/2017/09/societys-focus-on-curing-disability-can-harm-disabled-people/ September 13, 2017. Wentzel J. Researchers Discuss Ethical Considerations of 'Curing' Disabilities. Vanderbilt University https://news.vanderbilt.edu/2011/04/14/curing-disabilities-ethics/ April 14, 2011

[48] Among others, see Mercurio MR. The Aftermath of Baby Doe and the Evolution of Newborn Intensive Care. Georgia State University Law Review 2009;**25**(4): 835-863; White M. The End at the Beginning. The Ochsner Journal 2011;**11**(4): 309-316; and Annas Ag. The Case of Baby Jane Doe: Child Abuse or Unlawful Federal Intervention. American Journal of Public Health. 1984;**74**(7): 727-729

[49] Among others, see Wong A, ed. Resistance & Hope: Essays by Disabled People, Crip Wisdom for the People. Disability Visibility Project/https://wordpress.org, 2018; ADAPT, Free Our People, https://adapt.org (2019); Shakespeare, Disability, 48-50

[50] Ouellette A. Bioethics and Disability: Toward a Disability-Conscious Bioethics. Cambridge University Press, New York; 2011. p. 363-364.

Section 3

Models in Healthcare

Chapter 6

For a Model of Revision, Assistance and Care of Identities

Federico D'Angiolillo

Abstract

The global crisis scenario has highlighted the weaknesses of advanced personal assistance and care systems, based on the absolute primacy of technical knowledge. Almost all health organizations have been challenged by the new Coronavirus. The universal system because it is realistically unable to reach everyone efficiently and effectively. The private model, albeit moderated by intentions of global care, because it is onerous and, in fact, not very inclusive. This study, without any pretense of completeness, thanks to an examination of the most well-known documents published by the organizations for the promotion of human health, both EU and international, highlights the essential aspects and purposes of some of the main models of health care, also identifying the critical issues and the remedies prepared. The main purpose of the text is to highlight and reflect on possible alternative solutions to the current strategies to combat the pandemic, implemented by the states. The probable contributing causes that have contributed to the spread of the new coronavirus and its variants globally and that have their roots in now dated issues are then analyzed. The lesson that the Pandemic teaches us is that "no one is saved alone" and that the problems of each family, social, national etc., represent the problems of everyone. The document concludes in the sense that, only through a new approach to individual and collective health care, marked by greater solidarity and respect for individual, specific identities and frailties, starting from those "hidden" in society (adolescents, elderly, of handicaps, immigrants, etc.) it will be possible to promote welfare systems that are more attentive to the needs imposed by the challenges of globalization and therefore really more effective, economical and efficient, and therefore more humane.

Keywords: Pandemic, identity, health, fragility

1. Introduction

The global pandemic crisis shakes the situational equilibrium on which our realities are founded.

The pandemic shows the difficulties of modern social organizations to take care of the most disadvantaged and fragile people (elderly, people with disabilities, immigrants without residence permits, children and adolescents) [1] starting from each single identity [2].

As has been acutely observed [3], health promotion, understood according to the World Health Organization, as a state of physical, mental and social well-being, represents a productive factor of growth and development of the entire community.

But in recent years, governments attention has waned and produced a dismantling of the welfare state that has removed the social protection net of the most vulnerable people.

In this scenario, the pandemic caught us fundamentally unprepared, unable to oppose a reaction on the ground even of a cultural nature.

At the beginning of the crisis, the only serious collective defense tool was to take time by suspending fundamental freedoms, individual and collective, waiting for the vaccine.

On the now essential need for a common preventive strategy of defense against the danger of contamination by biological agents, and on the consequences produced by the pandemic, see, by way of example, the documents published by the World Commission for the Ethics of Scientific Knowledge and Technology Unesco (https://en.unesco.org/themes/ethicsscience-and-technology/comest) and the technical guides published by the WHO (https://www.who.int/emergencies/diseases/novel-coronavirus-2019) to which reference should be made.

Faced with today's crisis scenario, shared responses are necessarily urgent.

In this context, the use of vaccination represents, however, only one of the possible means of protection, not the only one, being by itself, unable to prevent the re-emergence of the elements favoring the current crisis situation on a global scale (which could be repeated) exposed (s) below:

- Loss of biodiversity - zoonoses (the pandemic of the new SARS-CoV 2 coronavirus is the result of the circulation of the virus from animals to humans [4] - have almost certainly been favored by pollution and habitat loss, by the creation of artificial, from the manipulation and trade of wild animals and more generally the destruction of biodiversity [5].

- Human interference in ecosystems - Human interference with the delicate balance of the climate system is taking place the risk of transmission of viral diseases [6].

- Abuse of the planet's natural resources - Need to change lifestyle, promoting a rebalancing of natural resources on the planet, reducing pollution factors (Co2) and strongly compressing the consumption of natural resources in developed countries.

If these factors did not immediately trigger the pandemic, they almost certainly have favored and, in some cases, accelerated its propagation with the consequences in terms of the cost of human, social and economic lives, which we know well.

For these reasons, the task of this chapter will be to investigate the factors that have contributed to the current pandemic crisis without neglecting the teaching that it brings with it: that of a great opportunity for rebirth.

2. A society of situational balances

The current pandemic crisis has shaken the balance of our life from the ground up. It has been estimated, in fact, according to a negative type 2 scenario, [7] that not before the end of 2020 the cost in terms of loss of human life in the United States alone would have reached 300,000 units. Unfortunately at the time of this writing the number of deaths it has exceeded, in the US, the 500,000 units, while globally far exceeded 3,000,000 deaths, [8] and is a number destined to grow.

At the end of the pandemic, the victims of Covid-19 - currently not quantifiable even through the use of the most advanced artificial intelligence algorithms - will be underestimated, even in the worst forecasts, with a large margin of deviation from reality.

Faced with such a scenario, it is therefore necessary to take a step back, questioning the reasons for the current social, economic and human vulnerability or at least explaining the anthropological and relational reasons that have given rise to this condition of generalized weakness.

In analogy with the great epidemics of the past, [9] the contagion has spread from east to west, affecting all the countries of the world.

The long and, unfortunately, cyclical wave of the pandemic (we are now in the 3rd wave) has changed our lifestyles, social habits, logical and relational patterns, both from the working point of view (employer-worker), but also personal-family (parents-spouses-children), behavioral, affective, personal-individual (man-woman).

A reflection is therefore required on the individual specific areas in which the pandemic has affected, in order to highlight not only the critical profiles, but also any positive aspects that this state of emergency has triggered.

3. Face-to-face work is not always good

Due to the pandemic, the massive use of new technologies made it possible not to suspend certain types of work services (for which it was not essential the physical presence of workers), which paradoxically appear to have strengthened in terms of organization and performance.

In particular, the spread of smart-working on a global scale that has been shown to be a useful, economical and sustainable remedy from the point of view of impact on the environment. Not only that, in the time of the restrictions imposed by the pandemic, it has also represented a formidable resource for employers who have achieved equal (if not higher) productivity results with less and better allocation of human and financial resources.

Remote work has made it possible to reduce physical and temporal distances by ensuring a large margin of time for the care and attention to fundamental relational interests (family, emotional and friends), but also to improve the quality of life and, in some cases, to devote personal time to volunteering [10].

According to a survey conducted by the American channel CNBC in collaboration with the online survey tool Survey-Monkey, people said they were happier doing their work remotely and an increasing percentage said they wanted to continue doing it even after the pandemic [11].

In conclusion, the experience of remote work successfully experienced during the pandemic could turn into a formidable tool for the transformation and reorganization of the world of work and society in order to make them more simplified and humans, not only for some social categories (workers, consumers, savers, etc.) but for all citizens.

4. Family relationships at the time of Covid: the role of women

According to a study conducted by Eurostat, the Pandemic would have burdened the work of women in the social and family contexts in which it is required with greater responsibility and affliction. According to Eurostat, the restrictions deriving from the lockdown would have had a greater negative impact on the female component, determining in particular in domestic contexts, the conditions for a significant increase in physical and moral violence, a complication (unpaid) of family duties and responsibilities, an extensive decrease in the employment rate of women, exclusion from top decision-making processes due to of the pandemic.

In particular, in the European context [12], it was observed how "the pandemic has separated families and friends, disrupted everyday life and even endangered democracies. It has affected every aspect of our European way of life. The crisis, however, has not been felt uniformly by every individual in our societies. Income inequality, geographic location, age and, in particular, gender, have determined, separately but also together, the way how the crisis has affected and will continue to affect citizens.

Gender and sex dominated not only the clinical aspects of the COVID-19 pandemic, but also how we responded to it. From urgent and pressing issues such as domestic violence and an alarming male mortality rate, to more structural and fundamental questions about the perceived value of different roles in society, it has become clear that gender has been an essential aspect of this virus and of relative crisis.

The competent Commission for women's equality of the House of Commons [13] of the United Kingdom Parliament also reached similar conclusions, who denounce how the Pandemic has triggered a significant increase in gender inequality, with particular reference to the percentage of requests for use of the sick state, and benefits also expected due to the pandemic, an increase in domestic violence against women, an increase in situations of poverty and female unemployment, an imbalance in the weight of domestic work and the burden of responsibility between genders, the request for a greater state economic support, professional retraining and reintegration into the world of work by women.

In conclusion, both of these studies conclude that the economic and social weight caused by the restrictions introduced due to the pandemic was brought especially to women, who would therefore have paid a heavy bill, both in terms of physical and moral health, and in terms of job insecurity and worsening of unpaid domestic responsibility situations, with a general condition of weakness and suffering, not only economic, suffered.

More generally, however, it should be remembered that the female "world" does not constitute a homogeneous humus, for which appropriate and more in-depth analyzes on the effects of the pandemic should be carried out precisely starting from the individual specificities of race, culture, religion, social and economic position of each woman (child, adolescent, elderly, handicapped, etc.) of the consequent damages received [14]. At the time of writing, there are no specific investigations on these particular aspects.

5. Affective relationships: the relationship between genders

The emotional relationships have been strongly affected by Covid [15]. People have been afraid, both for internal and external reasons, by choosing to lock themselves up in homes or by limiting their relationships in the home. So many people have poured their affection on the web in search of virtual hugs and relationships. However, there are distinctions between states. In Europe there has been a balance of physical relationships and virtual entertainment, (but for example in China virtual relationships have prevailed over reality). And in the UK? According to a United Nations study [16], the rapid spread of Coronavirus, on a global scale, would have occurred more in men (50.9%) than in women (49.01%) and probably also sexually. In fact, the British charity for the fight against AIDS - Terrence Higgins Trust - has repeatedly highlighted an increase in family relationships (84%). The restrictions on individual freedom caused by the lockdown have in fact pushed government bodies to more effective and accessible communication on the risks of infection with the new coronavirus even through safer options for avoid the risk of contagion, suggesting, for example, to book free exams [17] or promoting rules of responsible conduct and suspension of personal contacts to avoid infections and not recommending the taking medications due to the risk of exposure to HIV [18].

In conclusion, Terrence Higgins Trust argued that "the best way to fight the Pandemic, even from an emotional point of view, was to take care of themselves and protect each other. The best defense against the virus is to stay home as long as possible, follow government advice on limiting social contacts, keep two meters away from other people when you go out, wash your hands regularly, isolate yourself if necessary, and take care of the most vulnerable and isolated people" [19].

In the studies cited, the isolation resulting from the lockdown has produced pejorative effects of stable coexistence relationships between genders, male and female, fueling tensions, misunderstandings, relationship difficulties that have resulted in physical or verbal litigation, violence and in many cases, requests of divorce.

The consequences of the physical and psychological stress produced by Covid restrictions have been extensively studied [20] and the results obtained suggest public decision makers to adopt a temporally limited approach, maintaining a clear and effective communication capable of anticipating the negative effects that would inevitably have occurred in the medium and long term on the health system.

6. The limits of social health organizations

Historically, social health promotion and protection organizations have been of two types: public and private. Public health organizations have represented a flagship of the way to do health care, first of all in those States where the person has been really allocated at the center of human health care and assistance systems (e.g. in European countries, in Japan, in Brazil etc.). However, these advanced public social systems have shown profiles of dubious financial sustainability due to the high payment costs (tax levy) and the operational difficulties of ensuring the essential levels of assistance on a territorial basis [21]. In particular, the major aspects criticalities [22] that emerged, for example in Italy, concern: the care and assistance of people with comorbidities and chronic diseases; the inability to manage and treat and share health information flows; the poor interoperability of databases; the methods of taking charge of the assisted persons; the relationship with local communities [23].

From the universal health coverage system, the private health coverage service (adopted in the USA) is distinguished, on the basis of which health services are not provided mainly by the public health system, but by private structures, financed with health policies paid by patients assisted.

In this second case, access to treatment is only possible for citizens who have taken out private insurance. While the State concentrates, more or less incisively, in assistance and care programs for frail people (eg Medicare; Medicaid). In reality, even this system is not stable and suffers from political and economic conditions having been strongly influenced by the alternation of well-known politicalgovernment events which, in particular in the USA, have compromised it stability [24].

7. Universal and global health care systems at the time of Covid

Healthcare systems are characterized by two performance purposes: universal (UCH) and global (GHS). In the first case, the universal health coverage system tends to guarantee quality health services to all citizens regardless of financial problems [25]. From the universal health care system (UHC), the systems of health organization with global mandate (GHS) are distinguished that are focused on the prevention, detection and response to threats to public health, in particular from infectious diseases (USA, Africa).

According to the health security assessment mechanism [26], both of these systems would be in crisis as a result of the Covid-19 Pandemic for the following reasons:

- inability of national health systems to fight the prevention of pandemics;
- little funding in the safety tests of biological risk prevention and control systems on a global scale;
- lack of support actions for countries with political instability and health insecurity;
- strategic and programmatic inadequacy of national health systems in responding to pandemics;
- lack of coordination and training between policy makers and professionals in the implementation of the actions envisaged by the International Health Regulations (IHR) [27]
- non-alignment of national health systems with international standards on the risk of epidemics on a global scale.

According to a study published in the *Lancet* [28], countries that have invested in measures to adapt to international health security and access to treatment have been more effective in tackling the pandemic (eg South Korea, Singapore, Taiwan, Thailand, Cambodia, Kerala, Veneto Region). The *Lancet* study concludes on the need for nation states to promote concrete support and health security policies and actions for the future thanks to global health care systems of a universal nature that are perfectly integrated and aligned with common indicators that ensure for all integration, financing, resilience, equity.

8. Health as a productive factor for growth and employment: three strategies compared: European, American, British: the Italian universal health system

In continental policies, investments in public health represent a productive factor of growth and social development. The health protection and promotion strategy is also a testing ground for federal and parliamentary governments due to the growing emergency situation that has imposed severe revision policies. Below are some examples of health organization strategies and plans/programs launched to overcome the current crisis situation.

8.1 European Union

In the strategic documents of growth of the European Union [29], the right to health is almost always considered a fundamental driving force for growth not only social and intergenerational, but also economic. The European Union has in fact argued that thanks to efficient and innovative systems of care and assistance for people of productive age, it would be possible to achieve the objectives of containing public spending, improving the health of EU citizens by protecting them from international and transnational health threats.

The European Union has therefore launched a program for sustainable and inclusive growth, based precisely on the promotion of individual and collective health articulated on four fundamental pillars:

a. development of common tools and mechanisms at EU level to address the lack of human and financial resources and facilitate the uptake of innovation in health care, in order to contribute to innovative and sustainable health systems;

b. better access to medical expertise and information concerning specific diseases including on a transnational scale and to develop shared solutions and guidelines to improve the quality of healthcare and patient safety in order to improve access to better and safer healthcare for European citizens;

c. identifying, disseminating and promoting the adoption of validated good practices for cost-effective prevention measures, addressing key risk factors, notably smoking, alcohol abuse and obesity, as well as HIV/AIDS, with a particular focus on the cross-border dimension, in order to prevent disease and promote good health;

d. developing common approaches and demonstrating their value in being better prepared and better coordinated in health emergencies in order to protect citizens from cross-border health threats.

The impact on a global scale of the current pandemic emergency has had an undoubted conditioning effect on the action strategies for health promotion within the European Union.

In particular, the European Parliament approved a new EU Regulation no. 2021/522 establishing a Union action program in the health sector (so-called EU4Health program) for the period 2021–2027 and repealing regulation (EU) no. 282/2014.

Under the EU4Health program, the European Union for the next seven years will implement actions to combat the main cross-border health threats by creating:

- reserves of medical supplies for crises;

- a pool of health personnel and experts who can be mobilized to respond to crises across the EU;

- increased surveillance of health threats;

- strengthen health systems so that they can address epidemics and long-term challenges by stimulating disease prevention and health promotion in an aging population;

- digital transformation of health systems;

- access to health care for vulnerable groups;

- make medicines and medical devices available and accessible;

- support the prudent and efficient use of antimicrobials as well as promote medical and pharmaceutical innovation and greener manufacturing.

According to an INAPP research study, [30] as part of the actions already undertaken by the European Union for the promotion of health, the new European program "EU4Health" would establish a different and more effective system of connection of health measures within the European states integrated, also from the outside, with the protection instruments adopted at a supranational level.

In the new scenario outlined by the "EU4Health" program, the European Commission will exercise a leading role in the implementation of the planned actions. In particular, the EU Commission which can go as far as the exercise of a delegated legislative power, implementing in detail the indicators of the progress of the program (Annex II of EU regulation no. 2021/522) in compliance with the principles of precaution, complementarity, consistency and solidarity.

8.2 United States

In the United States of America, the federal government has approved a health reform plan "The Biden-Harris plan to beat COVID-19" articulated in seven key points.

In practice, the US administration [31] has focused its action to protect public health with the preparation of a plan to overcome the pandemic crisis - in the United States alone there are about 579,000 deaths - which aims to:

1. to give a greater voice to science.

2. Ensure that public health decisions are made on the basis of information provided by public health professionals.

3. Promote trust, transparency, common purpose and accountability in the federal government.

4. Ensure that all Americans have access to regular, reliable, and free testing.

5. Solve the supply problems of personal protective equipment (PPE) forever.

6. Provide clear, consistent and evidence-based guidance on how communities should address the pandemic and the resources for schools, small businesses and families to cope.

7. Plan for the effective and equitable distribution of treatments and vaccines, because development is not enough if they are not distributed effectively.

8. Protect older Americans and high-risk people.

9. Rebuild and expand defenses to predict, prevent and mitigate pandemic threats, including those from China.

10. Implement the use of IPR in the population, allowing people affected by the medium and long-term effects of the COVID-19 disease not to be excluded from the affordable quality public health care system, similar to the private one.

In addition to the health strategy of the US plan (*Biden-Harris*) for the escape from the crisis, it should also be remembered the adoption of a specific directive on national security aimed at recovering the "Global leadership of the United States to strengthen the international response COVID-19 and promote global health security and biohazard preparedness" [32] as well as the numerous executive measures (presidential actions) adopted for the protection and response to the spread of the coronavirus.

8.3 United Kingdom

Similar actions to respond to the pandemic threat (in the United Kingdom there are more than 128,000 deaths due to Covid) have been undertaken by the British government (British Public Health Plan - PHE 2025) which identifies at least 10 strategic priorities for the next five years and that the National Health System (NHS) is committed to supporting. In particular, there is also agreement in the United Kingdom on the need for greater internal collaboration and participation by the competent British Department of Health, including with the European Union, and the rest of the world, for the prevention of future pandemic scenarios. However, health protection in the British recovery program is not only achieved through the functioning of the national public health system (NHS), but also through the promotion of social and economic conditions that allow people to enjoy a life of dignity, prosperity and aware of the risks associated with bad eating and behavioral habits, improper and unbalanced lifestyles. (sedentary lifestyle, alcohol consumption, smoking, lack of social relationships). In conclusion, the British public health strategic program for the next five years (PHE 2020–2025) intends to implement the set of knowledge and skills to support better physical and mental health in the population. Improve the organization and the quality of the work of decision-making processes. Use cutting-edge tools and techniques to increase organizational efficiency and propose new solutions and virtuous approaches to public health problems.

8.4 Italy

Italy was one of the first countries in the world to be affected by the pandemic and to have recorded the highest percentage index of the number of deaths in relation to the population (about 120,000 deaths due to Covid). Specifically, the health emergency has revealed many structural and organizational weaknesses in the response to the growing demand for care and healthcare following the pandemic [33].

It should be borne in mind that the organization of the public health service in Italy is entrusted to the Regions (while it is the responsibility of the State only to determine the essential levels of assistance), and for this reason, it shows evident territorial disparities in the provision of health services, in particular of specialist services, in the prevention and management of comorbidities and chronic diseases, in the activation of a coordinated response to pathologies resulting from climatic, environmental and local alterations, also through the use of modern information and digital technologies. To counter these deficits (while we are writing, the National Recovery and Resilience Plan - PNRR has been approved) [34]. Italy over the next five years will allocate 15.63 billion euros for the creation of proximity networks, structures and telemedicine for territorial health care, innovation and renewal of existing digital technological structures (FSE, LEA) as well as staff training.

In particular, by 2026 it is intended:

- Strengthen the national health service by aligning services to the needs of communities and patients (Community Homes)
- Strengthen local health facilities and services and home services (home as the first place of care)
- To develop telemedicine and to overcome the fragmentation and lack of homogeneity of the health services offered on the territory (telemedicine).
- Develop advanced telemedicine solutions to support home care (intermediate care).

In a nutshell, the Italian government, thanks to the "Next Generation EU" European aid program, intends not only to implement the standards of universal care and assistance currently in place, through the use of modern communication and digital information technologies, but also to enhance the network of medical and territorial assistance services, with the creation of 381 new community hospitals for short-term hospitalizations.

9. The consequences of the pandemic

If even today there are no unambiguous certainties about the real origin of the current global pandemic crisis [35] we are certainly able to indicate with an appreciable degree of reliability at least seven negative effects, direct and indirect, which have occurred at the level:

a. first of all, the easy spread of the coronavirus, even by air, in particular environmental conditions, therefore without any direct contact, which has led to severe restrictions and obligations of conduct; [36]

b. secondly, short-term symptoms (high temperature, persistent cough, loss of taste and smell) and long-term (fatigue, shortness of breath, chest pain and tightness, memory loss, insomnia, palpitations, dizziness, tingling, joint pain, depression, feeling unwell, fever rash) caused by the infection; [37]

c. thirdly, the high number of deaths: 3,236,104 (dates May 6, 2021); [38]

d. fourthly, the loss of personnel in the healthcare sector due to the coronavirus; [39].

e. fifthly, the considerable use of physical and financial resources [40] for the response to the pandemic emergency;

f. sixthly, the victims of gender-based violence caused by the pandemic; [41]

g. seventh, the loss of jobs and economic initiatives [42].

All these effects describe the high price caused by the pandemic, both in terms of human lives, and in terms of social and economic costs that will have to be incurred for reconstruction and which are borne by future generations [43].

10. The real reasons for the crisis

From the indicated list of negative consequences determined by the pandemic, it is clear that the global crises has made vulnerable almost all our certainties (life, health, work) and lifestyles [44], strongly questioning the agendas, projects, habits and priorities [45].

After all, as the Holy Father correctly pointed out, what really highlights the pandemic is "*the evident inability to act together. Despite being hyper-connected, there has been a fragmentation that has made it more difficult to solve the problems that affect us all. If someone thinks that it was just a matter of making what we were already doing work better, or that the only message is that we have to improve existing systems and rules, they are denying reality*". Faced with the pandemic, the Holy Father concludes, the fragility of world systems has highlighted that not everything is resolved with the freedom of the market and that, in addition to rehabilitating a healthy policy that is not subject to the dictates of finance, "*we must restore human dignity to center and on that pillar the alternative social structures we need must be built*" [46].

But this lack of unity on a global level is also reflected at the internal level of individual states. "Similarly," the Pope reflects, "*the organization of societies around the world is still far from clearly reflecting that women have exactly the same dignity and identical rights as men. In words certain things are affirmed, but decisions and reality shout another message. It is a fact that women who suffer from situations of exclusion, mistreatment and violence are doubly poor, because they often find themselves with less chance of defending their rights*"[47].

It is evident that if society is in imbalance, if it is difficult to affirm and promote the rights of every person to participate in the human consortium (women, children, elderly, disabled, etc.), all the more this imbalance will reflect at a global level, that is, in sharing and promoting effective development policies even for the most disadvantaged nations. This is why it is essential to rethink the ways of identifying places, tools and opportunities to bring out and affirm the rights of these "hidden" identities which, on the other hand, can contribute to the civil and democratic progress of every social structure [48].

11. For a possible way out

So, if we assume that the true stability of the civil consortium has not yet been achieved and indeed has worsened due to the pandemic, in this perspective they will have to be positively welcomed the three recommendations of the United Nations to overcome the global crisis (global health care needed; common socio-economic and humanitarian policies; a recovery plan that rewards gender equality) [49].

From an economic and employment point of view, particular attention will be paid, among other things, to the strategy of social dialogue (so-called fourth pillar of the ILO strategy) between governments and individual social partners, both at national and transnational level, for the identification of a conscious common strategy for resolving the crisis.

Finally, from the health point of view, in this phase of the pandemic, it is necessary to guarantee effective, efficient and resilient global immunization services, that is, accessible to all people through basic health care of a universal character [50].

In the near future, on the other hand, the international, national and supranational recommendations on prevention of zoonoses and arbovirosis [51] must be scrupulously followed and applied through the implementation of epidemiological intelligence actions, training and information for operators and citizens, prevention and control from the risk of exposure, contamination and transmission of biological agents.

In a globalized way, the measures and actions for the control and contrast of the pandemic must necessarily be shared and coordinated with strategic actions and agreements of States and communities of States that privilege the centrality of the human person on the economic interests of production [52].

We must be aware of the planetary significance of the factors that have conditioned this crisis, which require shared responses. In an overall logic, therefore no longer based on partisan interests, it will also be possible to face and defeat the next global crises that have their roots, in hindsight, in issues that have not been resolved for years, if not for centuries and which, here, they can only be mentioned in summary: the fight against hunger in the world; the cultural, economic and civil divide between rich and poor countries; the sharing of scientific and technological advances, the promotion of fundamental human rights, respect for nature and the environment.

In this direction, there has been a significant change in the steering strategy of the World Trade Organization (WTO) regarding a possible suspension of the so-called "TRIPS" agreements, which govern the protection of intellectual property rights under international customs law, in this case, resulting from the production of vaccines and treatments anti-Covid [53].

The dialogue initiated within the WTO, regarding a possible suspension of the TRIPS agreements, had a first favorable echo from various states, including the US Administration [54] and was the subject of particular attention also in the EU context on the occasion of the G20 World Health Summit, on 21 May 2021, held in Rome [55].

The hope is that, also thanks to this important international forum of the main world economies, this year under Italian leadership, common actions and policies can be implemented to fight the pandemic, founded on solid ethical foundations about the need for a common strategy to overcome this crisis [56].

12. Conclusions

As a consequence of the current global crisis scenario, this chapter has shown the characteristics, but also the criticalities, of the main and most advanced systems of care and the promotion of human health.

The examination of some of the most well-known Western health organizations has revealed a scenario of crisis that has sharpened as a result of the pandemic, which are no longer able to respond effectively and promptly to the new challenges caused by globalization.

For this reason almost all systems Western health care and protection are under review.

This is good. However, the new organizational structure will have to be more attentive to the real needs of society.

In some crucial areas of medical research, the interests of politics do not have to prevail and we need to give more weight to science.

The challenges of the next years, even in the health sector, therefore require better and more prudent skills protection and information on possible risks to the health and safety of any personal identity, in particular children, the elderly, the disabled, immigrants, pregnant women, with a spirit of greater solidarity [57].

History teaches us that, only thanks to integration and authentic respect for specific cultural, social and religious identities is it possible to create a generous and peaceful coexistence that nourishes conditions of shared well-being and development.

In order for all this to be concretely possible, however, it will also be necessary to treasure the teaching that the Holy Father gave to the current pandemic, which reminds us that:

"Where nature and, even more, persons are involved, another way of thinking is needed, one that can broaden our gaze and guide technology towards the service of a different model of development, more healthy, more human, more social and more integral" [58].

Author details

Federico D'Angiolillo[#]
National Institute of Statistics, Rome, Italy

*Address all correspondence to: federico.dangiolillo@gmx.fr

The opinions expressed in this work constitute, exclusively, the result of personal elaborations, therefore they cannot bind - in any way - the Administration to which they belong.

References

[1] Presidency of the Council of Ministers - CNB "Covid-19: Public Health, individual freedom, social solidarity" - May 28, 2020 (cf. http://bioetica.governo.it/media/4112/p137_2020_covid-19-salute-pubblica-libertaindividuale-solidarieta-sociale_it.pdf)

[2] Pope Francis, Encyclical letter «All Brothers» - 3 October 2020

[3] European Commission "Questions and Answers on the New EU4Health Program" - May 28, 2020 (cf. https://ec.europa.eu/health/home_en)

[4] World Health Organization "WHO-convened global study of origins of SARS-CoV-2: China Part" - 14 January - 10 February 2021 (Cf. https://www.who.int/publications/i/item/who-convened-global-study-of-origins-of-sarscov-2-china-part)

[5] WWF - The worst epidemics in recent history, March 15, 2020 (Cf. https://www.wwf.it/pandemie_e_distruzione_degli_ecosistemi/)

[6] ISPRA - Cities and the challenge of climate change - Quality of the urban environment - 10th Report, 2014 Edition (cf. https://www.isprambiente.gov.it/files/pubblicazioni/statoambiente/FocussuLecittelasfidadeicambiamenticlimatici.pdf)

[7] McKinsey & Company Report "When the covid 19 Pandemic will be over" - September 2020 (Cf. https://www.mckinsey.com/industries/healthcare-systems-and-services/ourinsights/when-will-the-covid-19-pandemic-end)

[8] The Millennium Project Covid Scenarios Team "Three futures of the covid-19 Pandemic in the United States, January 1, 2022", October 2020 (Cf. http://www.millenniumproject.org/wp-content/uploads/2020/10/Covid-Scenarios-Full-Report.pdf)

[9] J. Charles Sournia, 1993 - Epidemics in Treccani, Encyclopedia of Social Sciences;

[10] De Masi Domenico - Smart working. The revolution of intelligent work - 22 October 2020 - Marsilio Editore.

[11] Laura Wronski - CNBC - SurveyMonkey Workforce Happiness, May 2020

[12] Committee on Women's Rights and Gender Equality - Annual Report to the European Parliament - 5 March 2020;

[13] House of Commons Women and Equalities Committee - «Unequal impact? Coronavirus and the gendered economic impact", 26 January 2021;

[14] European Parliament "The gendered impact of the COVID-19 crisis and post-crisis period», September 2020;

[15] IPSOS "Italy during coronavirus-Relationships among people, loved ones and sexuality", April 16, 2020;

[16] UN Women «In Focus: Gender equality matters in COVID-19 response», 26 June 2020;

[17] Gov. UK "Guidance Coronavirus (COVID-19): getting tested", 15 April 2020

[18] Terrence Higgins Trust "Don't hook up during the COVID-19 lockdown", March 25, 2020;

[19] Terrence Higgins Trust «Coronavirus COVID-19», 25 June 2020;

[20] Samantha K Brooks, Rebecca K Webster, Louise E Smith, Lisa Woodland,

Prof Simon Wessely, Prof Neil Greenberg, FRCPsyc et al. Show all authors "The psychological impact of quarantine and how to reduce it: rapid review of the evidence", THE LANCET, Published: February 26, 2020 (Cf. https://www.thelancet.com/journals/lancet/article/PIIS01406736(20)30460-8/fulltext)

[21] Istat, Annual Report 2020;

[22] QuotidianoSanita.it «National Chronicity Plan: critical issues and solutions at the center of the V Conference "Health and local communities" of Active Citizenship» Rome, 02 March 2019;

[23] QuotidianoSanita.it «National Chronicity Plan: critical issues and solutions at the center of the V Conference "Health and local communities" of Active Citizenship» Rome, 02 March 2019;

[24] Kaiser Family Foundation "Filling the Coverage Gap: Policy Options and Considerations", 22 April 2021 (Cf. https://www.kff.org/medicaid/issue-brief/filling-thecoverage-gap-policy-options-and-considerations/)

[25] World Health Organization "Universal health coverage (UHC)", 1 April 2021

[26] GHS Index - Global Health Security Index (https://www.ghsindex.org/)

[27] World Health Organization "International Health Regulations", 2005.

[28] Arush Lal, Ngozi A Erondu, David L Heymann, Githinji Gitahi, Robert Yates «Fragmented health systems in COVID-19: rectifying the misalignment between global health security and universal health coverage», THE LANCET, dec. 1, 2020 (cf. https://www.thelancet.com/action/showPdf?pi i=S0140-6736%2820%2932228-5)

[29] Europe 2020: the European Union strategy for growth and jobs, 07 February 2017;

[30] Alfredo Rizzo INAPP «The new centrality of European health policies in the light of the "EU4Health" discipline» ARS, 10 April 2021;

[31] The White House "COVID-19 - The Biden-Harris plan to beat COVID-19", January 21, 2021;

[32] The White House "National Security Memorandum on United States Global Leadership to Strengthen the International COVID-19 Response and to Advance Global Health Security and Biological Preparedness", 21 January 2021;

[33] European Commission, OEDC Report 2019 - State of Health in the EU Italy, Health Profile 2019;

[34] European Commission, Recovery plan for Europe, 1 January 2021;

[35] Amy Maxmen, "WHO report into COVID pandemic origins zeroes in on animal markets, not labs" Nature, 30 March 2021 (Cf. https://www.nature.com/articles/d41586021-00865-8)

[36] Erin Bromage «The Risks - Know Them - Avoid Them» 6 May 2021 (Cf. https://www.erinbromage.com/post/the-risks-know-them-avoid-them)

[37] United Kingdom National Health Service «Symptoms of coronavirus (COVID-19)», 29 April 2021;

[38] World Health Organization (Covid-19), Dashboard, 06 May 2021;

[39] Amnesty International - Italy «7000 health workers died due to Covid-19», 3 September 2020; Healthcare Panorama - Information & analysis of welfare systems «Covid19 contagion, Nurses are the most exposed category of health workers», 09 March 2021;

[40] The White House, 21 January 2021 (source quoted)

[41] United Nations Women "The Shadow Pandemic: Violence against women during COVID-19", March 20, 2020;

[42] ILO - Rome «COVID-19 and the world of work», 25 January 2021 (Cf. https://www.ilo.org/rome/approfondimenti/WCMS_739996/lang--it/index.htm)

[43] David Cutler - Lawrence Summers "The COVID-19 Pandemic and the $ 16 Trillion Virus" JAMA, 12 October 2020 (cf. https://www.hks.harvard.edu/centers/mrcbg/programs/growthpolicy/covid-19pandemic-and-16-trillion-virus-david-cutler-and)

[44] NHS –UK «Social distancing and changes to everyday life», 06 May 2021

[45] Francis, Encyclical Letter «Fratelli Tutti - On Fraternity and Social Friendship», 03 October 2020

[46] Francis, «Extraordinary moment of prayer in times of epidemic», Churchyard of St. Peter's Basilica March 27, 2020.

[47] Francis, Encyclical Letter «Fratelli Tutti - On Fraternity and Social Friendship», 03 October 2020

[48] Francis, "Message of the Holy Father Francis for the International Day of Persons with Disabilities, 03 December 2020".

[49] United Nations – Covid 19 Response, source quoted (cf. https://www.un.org/en/coronavirus/UN-response); European Institute Gender Equality «Gender Equality Index- 2020» (cf. https://eige.europa.eu/gender-equality-index/2020)

[50] WHO - Immunization Agenda 2030: A Global Strategy to Leave No One Behind (cf. https://www.immunizationagenda2030.org/)

[51] CDC - OneHealth «Prioritizing zoonotic diseases using a One Health approach», 03 February, 2020 (Cf. https://www.cdc.gov/onehealth/what-we-do/index.html); WHO «Global vector control response: an integrated approach for the control of vector-borne diseases», 31 May 2017 (cf. https://apps.who.int/iris/handle/10665/275708) ECDC – European Union «ECDC activities on epidemic intelligence and outbreak response» 06 May 2021 (Cf. https://www.ecdc.europa.eu/en/about-us/what-we-do/ecdc-activities-epidemicintelligence-and-outbreak-response)

[52] People's Vaccine Alliance, 06 May 2021 (cf. https://peoplesvaccine.org/)

[53] The White House «Press Briefing by Press Secretary Jen Psaki and Secretary of Agriculture Tom Vilsack, May 5, 2021» (cf. https://www.whitehouse.gov/briefing-room/press-briefings/2021/05/05/press-briefing-by-press-secretary-jen-psaki-and-secretary-of-agriculture-tom-vilsack-may-5-2021/)

[54] World Trade Organization «DG Okonjo-Iweala underlines urgent need to address equitable access to vaccines» (cf. https://www.wto.org/english/news_e/news21_e/gc_05may21_e.htm)

[55] European Union «Global Health Summit» (cf.https://global-health-summit.europa.eu/index_en)

[56] G20 – Italia 2021 «Priorities»(c.f. https://www.g20.org/italian-g20-presidency/priorities.html)

[57] Francis, «Ecumenical and Interreligious meeting with young peaople» Adress of His Holiness, Skopie, 7 May 2019

[58] Francis, «Message of the Holy Father Francis to the participants in the "European House- Ambrosetti" forum, 4-5 September 2020

Chapter 7

COVID-19 Myths and Music Advocacy in Nigeria

Oludayo Tade

Abstract

Myth mongering constitutes major impediment to the fight against COVID-19 and adherence to COVID-19 safety protocols in Nigeria. Against this background, this chapter analyses three COVID-19 advocacy songs to unpack how lyrics were used to neutralise myths and articulates adherence to COVID-19 preventive protocols. To burst the myths that the COVID-19 is a disease of the rich and the aged, the lyrics indicated that the virus does not respect social status or spare any age group. The songs contributed to advocacy by preaching adherence to COVID-19 safety protocols to be safe and survive the 'plague'. The paper stresses the importance of incorporating religious institutions, particularly music evangelists, in the fight against pandemics and other health crisis.

Keywords: COVID-19, music, advocacy, health seeking behaviour, Pandemic

1. Introduction

This paper examines representations of myth in religious songs and how they contribute to the fight against COVID-19 pandemic in southwest Nigeria. Myths spreading during COVID-19 pandemic have been costly in getting effective policy intervention outcomes through the propagation of deliberate falsehood and unscientific conspiracy theories about the cause, care and consequences of taking certain actions. This is not the first time myths on health emergency is formed and shared. Wilkinson and Leach [1] examined myths during Ebola health epidemic and showed how myths and assumptions surrounding Ebola led to wrong responses. Most of the myths, such as drinking and bathing with salt water during the Ebola outbreak were found to be untrue.[1] Indeed, people were found to be hiding their sick people owing to cremation policy and non-engagement of communities to adopt creative approaches to burial practices [3].

While some studies have established the negative consequences of myths, others see positive outcomes in myths. The negative effects of myths have been further espoused by scholars who opined that myths (pandemic myths) can escalate fear, create panic and contribute to stigmatisation [4]. Mintzerberg [5] examined six myths within the health system in developed countries and found that statements claiming that health systems are failing are untrue. Kar et al. [6] emphasised the need to convey proper health information or scientific evidence to people to reduce fear and anxiety. Myths have their own value in helping people understand why things are the way they are [7]. To them, "in other contexts they are source of

[1] For more on behaviour response to Ebola, see Ogoina et al. [2].

IntechOpen

comfort, and help people explain where they come from, why the world is as it is, and why things are the way they are. However, they should have no such privilege place in evidence-informed public health".

The responses of people to the pandemic may have to do partly or wholly with the disseminated and consumed information [8]. As people transit from the physical to the virtual mode as one of the preventive protocols to reduce physical contact and enhance social distancing and prevent the spread of the COVID-19 virus, information sources became uncoordinated as people are bombarded with information online spreading myths and unfounded lies on the pandemic. The information systems and media reportage of the pandemic bombards the public with volumes of information to mine and make sense of. As a result, available information through different sources might be interpreted differently because people may not have the capacity to distinguish between the 'truth' and what is made available to them. Steps taken to curtail and contain COVID-19 such as lockdown measure have impact on the economy, education and mental health of the people [9]. Indeed, World Health Organisation has said the world is fighting both infodemic and pandemic [10]. Indeed, the World Health Organisation [11] listed some common myths associated with COVID-19 in a bid to burst them. The global health body maintained that the virus spreads in all types of humid environment to counter the myth that it spreads faster in cold weather than in hot or humid environment. The organisation stated that there is no scientific evidence to support the claim that mosquitoes spread COVID-19; that regular washing of hands and not hand driers kill COVID-19; that consumption of alcohol cannot kill COVID-19 but decreases the immunity of the body while increasing body vulnerability to the disease if infected and interfere in treatment procedure. That the claim that COVID-19 affects older people and cannot affect younger people is not true because the virus affects people of all ages, sex, race and only good hygiene is prescribed to prevent contraction. Even in Nigeria, the virus has killed all age categories. The negative effects of myth mongering were recorded during Ebola and Polio vaccination health crises. Myths mongering affected behavioural responses. People were asked to take salt or bath with it to prevent or cure Ebola. In Northern Nigeria myth also caused the boycott of polio vaccination [2, 12]. This present paper contributes to scholarship on myths and pandemics by examining the representations of myth in religious songs and how the advocacy content of the songs contributes to the fight against COVID-19 pandemic in southwest Nigeria.

2. Music, health and wellbeing

Music engages people and appeals to their moods, conditions and fear. It is a communicative act which has the power of reordering social conditions. Music has the power to reduce feelings of isolation and loneliness [13] and reduce tension [14]. There is a growing interest among scholars in examining the intersection or functionality of music in aiding health and well-being [15]. It follows therefore that listening to music is a self-medicated therapy which regulates and maintains health and sustain wellbeing. It works more if preferred music is available to patients. According to MacDonald et al. [16] listening to the right music is crucial to health and wellbeing.

Music has therapeutic value in enhancing recovery from ill-health and has been used in treating the problem of love sickness. Batt-Rawden [17] averred that "music as a powerful force to be used as an intervention in curing illnesses and disease". It is deployed to create happiness, harmony and engender togetherness and as Gouk [18] puts it, music cures illnesses or diseases in healing ceremonies. It is even said that through music therapy, the entire gamut of transformational healing that a patient is in need of is fully served [19]. Indeed, it can be used to cope with depression feelings and emotions and has agency to maintain orderliness [20, 21]. It follows therefore that music enables appreciation of social conditions, social realities and personal encounters and once place within it. Music sociology unpacks the selected music can be used as self-therapy [22]. It helps in coping with life's crisis such as presented by COVID-19 pandemic. Thompson [23] noted that music has a special appeal that makes people feel good even in times of crisis. According to him, music education has been used in China and Vietnam to educate people on staying safe:

> *The universal appeal of music as the go-to "feel good" experience in everyday life has been reported by news services covering the personal impact of social distancing and isolation. In China, dance parties have taken place both in make-shift hospitals and online. COVID-19 also has its own hit songs emerging, with tunes written to educate the public about hand washing and hygiene. A public education song from Vietnam evolved into an online dance challenge where people choreographed dance moves to help us wash our hands more effectively. People have created lists of 20- second songs to accompany our handwashing along with lists of "top albums to listen to" while in isolation. (23: 197)*

To provide meaning and education, COVID-19 songs were composed and released by Christian evangelists tapping into myths, calming anxieties and raising hopes through adherence to safety protocols. These songs provide interpretive understanding of the cause, effect and articulate adoption of preventive behaviours to check the spread of the pandemic in Nigeria. Through this, music becomes a social act which orientates and facilitates social action in relation to the pandemic. This study therefore unveiled the contributions of music evangelists to the fight against COVID-19 pandemic. It shows that music evangelists have agency to use their platform to preach conformist behaviours. This can be harnessed as vehicle to propagate safety protocols aimed at fighting health crisis such as presented by COVID-19.

3. Methods

To fulfil the research goals of this paper, three songs composed by popular Christian musicians were purposively selected for the study. These are Esther Igbekele's 'Only one plague', Yinka Ayefele's 'COVID-19 Prayers' and Ebenezer Obey's Coronavirus *alejo lo je o* song on COVID-19. These songs were released during lockdown in 2020 in Nigeria when different conspiracy theories were flying around. The songs selected because they provide understanding of the COVID-19 pandemic, myths and neutralisation, and advocacy for adherence to COVID-19 protocols to prevent the spread and contraction of the deadly virus. Two of the songs were rendered in Yoruba language while Esther Igbekele's song was in English. Due to this, I translated the Yoruba lyrics to English while ensuring that the social constructed meanings in Yoruba are not lost in the translated English. The data was thematically analysed in line with lyrical patterns which emerged from the songs.

4. Nature of COVID-19 myths

Several myths and misinformation have been shared since the confirmation of first COVID-19 index case in Nigeria on February 27, 2020. People thought it was a hoax. This unbelief may not be unconnected with the fact that Nigeria, and Africa have recorded lowest form of COVID-19 infection and the number of human fatalities have been minimal when compared to countries in the global north. Some believed that COVID-19 was a scam and that the leadership of Nigeria were only creating fear in people to make money from donations coming from international organisations and donor agencies [24].

However, available data from the World Health Organisation (WHO) does not support such positions about COVID-19. The numbers of those who contracted the virus, those that recovered and discharged as well as those who died are facts that confirm the realness of COVID-19.[2]

There was also a myth that COVID-19 will not survive in hot weather in an environment like Nigeria. This myth thrived because of the lower rate of fatalities recorded in Africa when compared to high death figures recorded in the UK and US. There was also a myth that the pandemic affects old people and not young people. How then did the lyrical representations of COVID-19 help in bursting some of these myths? What advocacy did the music do using their platform? In what follows, I analyse how the lyrical constructs burst myths on age and COVID-19, social class and COVID-19.

5. Bursting the age and social class myths

Myths surrounding what age and social class of people at the risk of COVID-19 spread across Nigeria. It was initially stated that COVID-19 does not affect the young but only the old. The myth surrounding social class was rife due to news reportage of the deaths of high profile Nigerians resulting from COVID-19 complications. It was thought that COVID-19 was the disease of the rich and not that of the poor (*aisan olowo ni, kii se ti mekunnu*). These myths created a false sense of safety for the poor and young with implications on them to lower their guard and not observe the COVID-19 safety protocols. This position was also strengthened by those who recovered from the virus. Doyin Okupe is a medical doctor and former media adviser to former President Olusegun Obasanjo. After recovering from the virus, Okupe claimed that more elite and rich people were dying because of their lifestyle. According to him, the poor are more exposed to the sun while the rich are not and consequently deficient in Vitamin D [25]. His status as a medical doctor complicated the issue and deepened the narrative about the high vulnerability of the rich. This accounted for why some people of low social-economic class took safety protocols for granted.

To burst this myth, evangelist Ebenezer Obey Fabiyi in the song, Coronavirus *alejo lo je o*[3] informs his listeners that COVID-19 does not discriminate like human beings and would inflict the same harm on all humans that fail to be on guard. To him, COVID-19 is not a respecter of social class (rich or poor) or age (young or old). Through his lyrical construct, Obey warns against the harmful consequences of listening to such myths and not taking positive steps to protect oneself and significant others from contracting COVID-19.This is expressed early in the song:

[2] As at 29-04-2021, a total number of 1, 912,628 samples have been tested. Out of these, there were 164, 933 confirmed cases, 155, 021 discharged, 7,909 active cases while 2,063 died of the COVID-19 complications. More data is supplied by the National Centre for Disease Control (NCDC): https://covid19.ncdc.gov.ng/.

[3] Coronavirus you are visitor.

Coronavirus you are a visitor,
it is killing people around the world,
it does not know the rich and does not spare the poor,
it does not spare the young or old
(Ebenezer Obey/Coronavirus alejo lo je o/2020)

Despite neutralisation age and social class myths through lyrics, these music evangelists joined the advocacy to embrace the scientifically proven preventive protocols. They amplified these safety protocols in their lyrics and enjoined their audiences to adhere to them to be able to survive the 'plague'.

6. "We should listen to what government is saying": music advocacy for preventive behaviour against COVID-19

Apart from neutralising myths and promoting scientific truths about the COVID-19, the lyrics also embed in them, advocacy on how to stay safe as the pandemic rages on. Prior to the release of the songs, Nigeria had declared lockdown and reeled out policy which must be adhered to while conducting themselves in the public as it eases lockdown. The Coronavirus Disease Health Protection Regulation 2021 signed by President Muhammadu Buhari placed restrictions on gatherings, mandated use of face covering in public, ordered 50% capacity reduction of worshippers in religious spaces and ban sharing of items among others. In Banks and schools, clean environment and provision of hand sanitisers are to be observed among others. Transporters are to reduce passengers and engage in frequent cleaning and disinfection of their vehicles regularly[4].

Notwithstanding these provisions, enforcement was weak and non-adherence was the outcome. To advocate adherence and observance of the safety protocols, lyrics of the COVID-19 songs were focussed on this aspect to the safety of all. *Coronavirus alejo lo je o* counselled people to obey the law and take heed to what God has instructed to prevent them from contracting COVID-19. Specifically, Ebenezer Obey's song restated what is required of Nigerians to do which aligns with the safety protocols of government. He reasons that if there is adherence, COVID-19 will not 'visit their households'. By preaching the benefits of adherence, the musician uses religious songs to encourage adoption of safety behaviour. This advocacy is captured in the lines below:

Coro oo, corona, do not branch in my household,
Let us follow the law and listen to what Government is saying,
Stay in your house, don't entertain visitors
We should wash our hands regularly
We should wash our hands regularly with water and soap, we should wash it regularly,
We should use sanitiser to wash our hands, we should cover our nose with facemasks,
we should avoid being close to one another (physical and social distancing), and not entertain visitors,
we should listen to what government is saying (Ebenezer Obey/Coronavirus alejo lo je o/2020)

[4] For more details on this health Regulation, see: https://covid19.ncdc.gov.ng/media/files/COVIDResponseMarch1.pdf.

Specifically, Ebenezer Obey picks specific audiences for his message and targets commercial transporters. He counsels against carrying overload to allow for reasonable distancing on board. By incorporating transportation sector, Obey recognises the importance of human mobility in the spread of COVID-19. Since movement cannot be halted, observing safety protocols is the way out of the woods:

We should maintain reasonable distance inside vehicle/transportation bus,
We should maintain distancing inside transport bus,
You driver must not carry overload, we should listen and obey government on preventing coronavirus,
The Okada rider (commercial motorcylists) should not carry more than one person, it is dangerous, we should heed the counsel of government,
Corona pack your load, it should not branch in my house, I have covered my nose, let us embrace cleanliness
(Ebenezer Obey/Coronavirus alejo lo je o/2020)

The above lyrics captured other aspects of the protocols including cleanliness and covering of nose. To Ebenezer Obey therefore; Corona virus will not branch in the house of whoever upholds this instruction from government.

In COVID-19 Prayer, like Corona virus alejo lo je o, Ayefele mentioned the potential outcome of those who refused to observe advisory protocol. Such a person, he said would have to yield all his life's investment to be enjoyed by another person since contracting COVID-19 may result into death although there are chances of recovering. While leaving the door open for a change of mind, Ayefele tells his listeners and fans not to shake hands, hug and should wash their hands with soap and water to be able to stay alive to enjoy the fruits of their labour.

Hope you have heard that they said we should not shake hands and not hug?
they also said we should wash our hands with soap and water and should be hygienic ...
Oh Lord, let us not contract corona so that a stranger will not reap the fruits of our labour ...
(Ayefele/COVID-19 prayer/2020)

In *Just one plague* Esther Igbekele calls attention to the need to observe social distancing and temporarily stop hand shaking and holding each other until coronavirus 2019 is defeated. *Just one plague* portrays the terrifying impact COVID-19 has had on the global economy stressing the need to obey the social and physical distancing protocol to prevent the 'only one plague' which has silenced even political unrest and protest which engulfed the world before its arrival. This is captured below:

Don't shake hands again,
don't hold each other again,
step away a meter just for only one plague.
Before the plague,
nations were threatening nations for war,
there was war in Syria,
there was revolt in Iran,
there was crisis in Turkey,
there were protests and political unrest,
bokoharam but when the plague surfaced,
everywhere was silenced. (Just one plague/Esther Igbekele/2020)

Despite their belief in divine intervention the three songs embraced scientific solutions and urged their followers and fans as well as the general populace to follow prescriptive guidelines to prevent the contraction and spread of COVID-19.

7. Conclusion

Against the background that studies are beginning to show the functions of music during times health crisis such as COVID-19 during lockdown [26–28], this paper examined the role of gospel music in the articulation of COVID-19 preventive protocols set out to check the spread and contraction of the pandemic in Nigeria. Three songs were purposively selected due to their lyrical contents which focussed on COVID-19 pandemic. We analysed their neutralisation of existing age and social class myths. Indeed the songs used the negative consequences (death and loss of one's investments to another person) of contracting the disease to show its realness. In advocating for adherence to COVID-19 protocols, the musicians deployed the fear and appeal approach to show the consequences of not observing the protocol against the benefit of observing the protocol.

Put together, the analysed songs articulated observance of safety protocols in the fight against COVID-19. Considering the cardinal role that religion plays in social construction of reality, it is recommended that music evangelists be incorporated in the future fight against health epidemic or pandemic and other social problems.

Author details

Oludayo Tade[1,2]

1 Department of Sociology, Victimology and Criminology Unit, University of Ibadan, Nigeria

2 Diaspora and Transnational Studies Unit, Institute of African Studies, University of Ibadan, Nigeria

*Address all correspondence to: dotad2003@yahoo.com

References

[1] Wilkinson Annie and Leach Melissa (2014). Briefing: ebola – myths, realities, and structural violence, African Affairs, 114/454, 136-148

[2] Ogoina D, Oyeyemi AS, Ayah O, Onabor AA, Midia A, Olomo WT, et al. (2016). Preparation and Response to the 2014 Ebola Virus Disease Epidemic in Nigeria The Experience of a Tertiary Hospital in Nigeria. PLoS ONE 11(10): e0165271. doi:10.1371/journal. pone.0165271 qhw.v8i0.20635

[3] Idowu, Abiodun Benjamin, Okafor Ifeoma Peace, Oridota Ezekiel Sofela and Okwor Tochi Joy (2020). Ebola virus disease in the eyes of a rural, agrarian community in Western Nigeria: a mixed method study. BMC Public Health https://doi.org/10.1186/ s12889-020-09441-7

[4] Ishmaev Georgy; Dennis Matthew & van den Hoven, M. Jeroen (2021). Ethics in the COVID-19 pandemic: myths, false dilemmas, and moral overload. Ethics and Information Technology, https:// doi.org/10.1007/s10676-020-09568-6

[5] Mintzberg H. (2018). Managing the Myths of Health Care. In: Adinolfi P., Borgonovi E. (eds) The Myths of Health Care. Springer, Cham. https://doi. org/10.1007/978-3-319-53600-2_1

[6] Kar Sujita Kumar, S. M. Yasir Arafat, Russell Kabir, Pawan Sharma, and Shailendra K. Saxena (2020). Coping with Mental Health Challenges During COVID-19, pp,199-213 In S. K. Saxena (ed.), Coronavirus Disease 2019 (COVID-19), Medical Virology: from Pathogenesis to Disease Control, https:// doi.org/10.1007/978-981-15-4814-7_16

[7] Viehbeck, Sarah M., Petticrew, Mark and Cummins Steven (2015). Old Myths, New Myths: Challenging Myths in Public Health. American Journal of Public Health, Framing Health Matters. Published online ahead of print February 25, 2015

[8] Ågerfalk Pär J , Kieran Conboy & Michael D Myers (2020). Information systems in the age of pandemics: COVID-19 and beyond, European Journal of Information Systems, 29:3, 203-207, DOI: 10.1080/0960085X.2020. 1771968

[9] Ågerfalk Pär J, Kieran Conboy & Michael D Myers (2020). Information systems in the age of pandemics: COVID-19 and beyond, European Journal of Information Systems, 29:3,203-3,207, DOI: 10.1080/0960085X. 2020.1771968

[10] Zarocostas, J. (2020). How to fight an infodemic. The Lancet, 395(10225), 676. https://doi.org/10.1016/S0140-6736(20) 30461-X

[11] World Health Organisation (WHO, 2020). Getting your workplace ready for COVID-19 (PDF). World Health Organization. 27 February 2020

[12] Jegede AS (2007). What led to the Nigerian boycott of the polio vaccination campaign? PLoS Med 4(3): e73. doi:10.1371/journal.pmed.0040073

[13] Clift, S., Hancox, G., Morrison, I., Hess, B., Kreutz, G. & Stewart, D. (2010). Choral singing and psychological wellbeing: Quantitative and qualitative findings from English choirs in a cross-national survey. Journal of Applied Arts and Health, *1*(1), 19-34

[14] Valentine, E. & Evans, C. (2001). The effects of solo singing, choral singing and swimming on mood and physiological indices. British Journal of Medical Psychology, *74*(1), 115-120

[15] Bunt, L. (1994). Music therapy: An art beyond words. London: Routledge.

[16] MacDonald, Raymond A. R. (2013). Music, health, and well-being: A review, The Arts in Psychotherapy 37 (2010) 301-310

[17] Batt-Rawden Kari Bjerke (2010). The benefits of self-selected music on health and well-being. International Journal of Qualitative Studies on Health and Well-being, 8:1, 20635, DOI: 10.3402/

[18] Gouk, P. (2000). Musical healing in cultural contexts. Aldershot: Ashgate

[19] Gagner-Tjellesen, D., Yurkovich, E. E., & Gragert, M. (2001). The use of music therapy and other ITNIs in acute care. Journal of Psychosocial Nursing, 39(10), 26-37, 52-53

[20] DeNora, T. (2000). Music in everyday life. Cambridge: Cambridge University Press

[21] Covington, H., & Crosby, C. (1997). Music therapy as a nursing intervention. Journal of Psychosocial Nursing, 35(3)

[22] Greasley, A. E., & Lamont, A. M. (2006). Music preferences in adulthood: Why do we like the music we do? In M. Baroni, M. Baroni, et al. (Eds.), Proceedings of the 9th International Conference on Music Perception & Cognition (ICNPC9). The Society for Music Perception and Cognition (SMPC) and European Society for the Cognitive Sciences of Music (ESCOM) Bologna, Italy, August 22-26, (p. 960)

[23] Thompson Grace (2020). Reflecting on relational needs in the context of a global health crisis, Nordic Journal of Music Therapy, 29:3, 197-199, DOI: 10.1080/08098131.2020.1746551

[24] Onapajo Akeem and Adebiyi Jelili (2020). 'COVID-19 is a Big Scam' Citizens' Distrust and the Challenge of Combating Coronavirus. The Republic, March 30, 2020. https://republic.com.ng/february-march-2020/distrust-nigeria-coronavirus/ accessed on 18/5/2021

[25] Obinna, Emelike (2021). Why prominent Nigerians are dying more than the poor. Business Day newspaper, January 17, 2021.https://businessday.ng/coronavirus/article/why-prominent-nigerians-are-dying-more-than-the-poor/ accessed on 18-5-2021

[26] Thompson Grace (2020) Reflecting on relational needs in the context of a global health crisis, Nordic Journal of Music Therapy, 29:3, 197-199, DOI: 10.1080/08098131.2020.1746551

[27] Chiu Remi (2020) Functions of Music Making Under Lockdown: A Trans-Historical Perspective Across Two Pandemics. doi: 10.3389/fpsyg.2020.616499

[28] Ziv Naomi and Hollander-Shabtai Revital (2021) Music and COVID-19: Changes in uses and emotional reaction to music under stay-at-home restrictions, Psychology of Music, 1-17 DOI: 10.1177/03057356211003326

Section 4

New Technologies

Chapter 8

Bioethics of Brain Computer Interfaces

Akram Jassim Jawad

Abstract

Nowadays, smart home devices have started to take a part in everything in our life, which mainly have been developed to consist from brain computer interface (BCI). In recent months, Neuralink BCI (1024-Electode) has been approved to be used by Food and Drug Administration (FDA) in the USA. That makes the ethical related studies have more attention to apply these devices and technologies in our daily life with more security. In this chapter, the ethical challenges of smart home systems that use BCI for personal monitoring, such as Neuralink Interfaces, have been reviewed, analyzed and discussed regarding the fundamental principles in 'Statement of Ethical Principles for the Engineering Profession' of the UK. Firstly, a brief introduction of Neuralink BCI technology and important applications in daily life were discussed with related ethics issues. Then, proposed solutions and recommendations for every situation have been introduced and discussed as well. The main proposed ways to address that are establishing and introducing the related laws and rules, technology development of security and safety, and educate for acceptance culture in the society.

Keywords: Brain Computer Interface (BCI), Engineering Ethics, Bioethics, smart home, Neuralink

1. Introduction

One of the most important and recent smart home devices for personal monitoring based on brain computer interface (BCI) is Neuralink BCI implanted that invented and produced by Elon Musk company "Neuralink" [1]. Recently, Neuralink BCI (1024-Electode) has been approved to be used by Food and Drug Administration (FDA) in the USA, which makes the ethical related studies have more attention to apply these devices and technologies in smart home applications in more security [2]. This device can help people with disabilities to communicate better and control devices by them brain thinking activity [3]. For example, this device can transmit the inner speech activity of the brain into an external speaker device for people who suffer speaking problems and illnesses such as locked-in syndrome LIS [4]. Moreover, this technology could produce a revolution in the treatment of different human brain related illnesses, for example epilepsy, schizophrenia, paralysis, and even brain injury. However, these devices could also make the society have social illness related with ethics issues such as inequalities, hackers, governments or people to manipulate and control other people, and many other social problems. The aim of this report is to discuss and write a personal reflection on the ethical challenges of smart home systems that use BCI for personal monitoring, such as Neuralink Interfaces, regarding the fundamental principles in 'Statement of Ethical Principles for the Engineering Profession'. Therefore, a brief introduction of Neuralink BCI technology and important applications in daily life were discussed with related ethics issues, and proposed solutions for every case as well.

2. Neuralink brain computer interfaces

Generally, the main components and steps in any BCI process were shown in **Figure 1**, which includes firstly an implanted array of electrodes to read brain signals, and then transition of that signals information into a receiver machine such as a computer. That BCI has the ability to read brain signals data and apply it in the control process for a specific activity [5]. However, in the implanted process of BCI device from Neuralink Company, about 3072 electrodes per array with 32 electrodes per bundle are inserted by robot. These electrodes give the ability to read and write the brain to allow humans to communicate and control many external devices [6]. Schematics of The implanted device in human, design and an experimental model

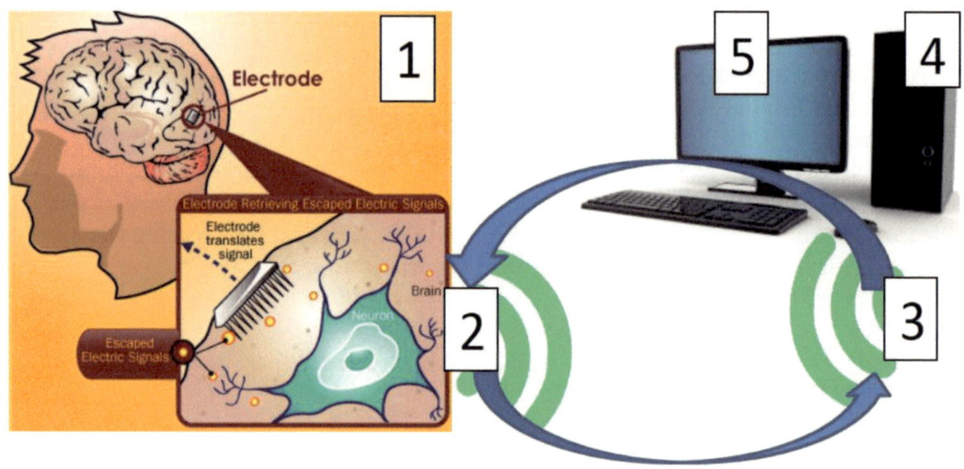

Figure 1.
An example illustration of a brain computer interface product that contain (1) implanted array of electrodes to read brain signals and (2) transition of that signals information into (3) a receiver machine such as computer (4) which has ability to read brain signals data (5) to apply it in control process for a specific activity, created by Microsoft office.

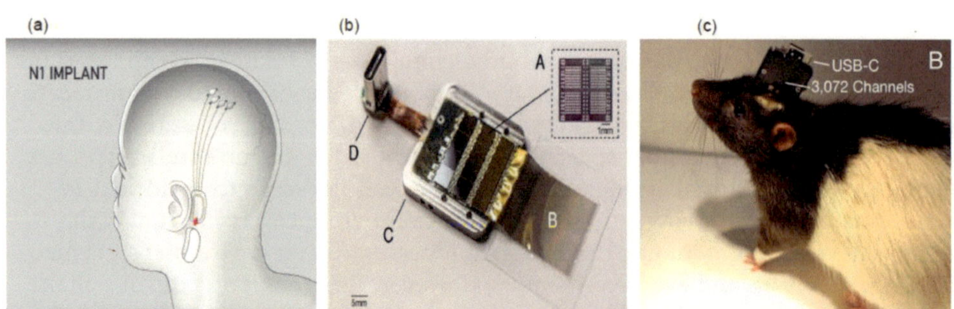

Figure 2.
(a) The BCI implanted device in human by Neuralink company. (b) BCI design with a USB-C interface. (c) Experimental of Neuralink BCI in a rat model [7], the authors have been emailed to get permission.

in a rat model of BCI and a USB-C interface by Neuralink company were shown in **Figure 2** a, b and c, respectively [7]. Consequently, there are a number of ethical issues that have to take in account when this device used in any application, such as health care or even in entertainment systems.

3. Ethical issues

One of the most related fundamental principles to Neuralink BCI technologies and devices in 'Statement of Ethical Principles for the Engineering Profession' is point two, which titled "Respect for life, law, the environment and public good". Their main points were summarized in **Table 1**, which clarify the comment on the application possibility of these limits in the targeted devices of this report. In the literature, it has been suggested that there are four different areas of ethical concern, which are privacy, agency, identity, and bias as well [8]. It has been mentioned that the governments have to establish regulations, rules, and laws depending on them public ethnicities, religious and socio-economic culture for them nations. **Figure 3** shows appications of BCI, and **Figure 4** represents a summarizing show which cover all possible ethics issues in BCI that proposed by Coin et al., (2020), which includes three main factors are physical, physiological and social actors [9]. The physical factor is just the user safety, the physiological factors are humanity and personhood, as well as autonomy. While the social factors are stigma and normality, responsibility and regulation, research ethics and informed consent, privacy and security, and justice [9].

As it has been mentioned, there are six groups of ethical challenges of BCI Neuralink's devices. The first one is the device has to be lawful and justified work, which means that the device has to be approved before its ethical issues will be considered. For example, Neurohype, which means publicity for some devices and

Regulations/laws	Possible Application in BCI products
Health and safety requirements	Low, and needs to rewriting and stablishing
Clarify the hazards	Needs to be more clear and specified
Lawful and justified work	Needs to be acceptable in local or national organization, such as the MHRA (Medicines and Healthcare products Regulatory Agency) in the UK
Physical and cyber security	Low, and needs to provide a technology with high safety
Personal information and intellectual property	Needs both establishing laws and invented technology to provide high safety

Table 1.
The main factors of point 2 in fundamental principles in the UK that were written in 'Statement of Ethical Principles for the Engineering Profession', which titled "Respect for life, law, the environment and public good", with comments on possible application for BCI products.

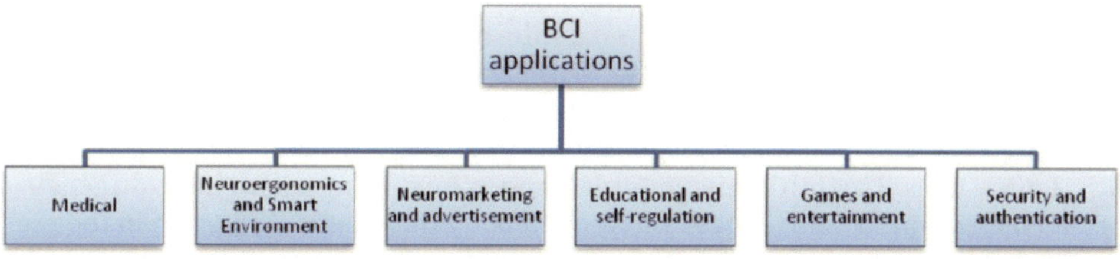

Figure 3.
Applications of BCI, created by Microsoft office.

Figure 4.
A summarizing show which cover all possible ethical issues in BCI.

technologies before they have been approved by scientist society, or even before they have been made, which are technologies false and exaggerated or even unsubstantiated claims [10]. The second challenge is to clarify the hazards and research ethics officially from the producer and the organizations. Where, spreading the culture and increasing the awareness regarding the BCI devices not just between the specialist scientist but also between public society, such as the EU and the UNESCO, play a significant role to solve the fair access of both technology and its ethics [11]. Therefore, the transparency of using these devices have to be highly clear to avoid the deception with or without intention, and to make and encourage equality and fairness in the society. The third issue is the physical safety. As Neuralink's devices and technology involves inserting thousands of electrodes into the brain surface, it uses by helping a robot to identify and avoid targeting neurons and blood vessels as possible [9]. However, this process may produce physical safety issues such as infections and distortions of neurons, tissues and blood vessels, or reactions of the immune system. Therefore, the developing and improving in these robot technologies are necessary to minimize the physical safety concerns.

Another point is the privacy that is related to personal information, and identity issues. As this device uses a USB interface, Wi-Fi or Bluetooth ways to transmit the data, which are likely to be hacked with or without permission, and as an implanted device that gives physical pain under some conditions as it has been mentioned before, together raising hybrid ethical concerns [10]. This could be a big issue in terms of safety considerations. As a result of that, the scientists and suppliers community should provide high safety that the access for these devices only under control of the users. For example if a person has done a crime and consequently the police wanted abscess to access to their device to do the investigation [12]. This police procedure may produce high ethical concerns of the personality data and privacy, which could be widely dangerous and harmful in non-democracy governments. Kellmeyer (2018) mentioned that the data protection of the brain have to be in the top of safety guidelines and protection regulations in BCI devices, to solve the important ethical issue in this technology that is personal data protection and mental privacy of users [5]. People should have high ability, freedom and the right to protect their private neural and mental information [8]. Also, it is clear that the data will be one of the main things that sells in what is called the black market, which needs to stablish new laws to organize and control that.

The agency, autonomy, and responsibility play a significant ethical concerns in BCI devices uses. This ethical issue could happen by sharing such a hybrid agency between these implanted devices and human body that makes limitation in the human autonomy range in different activities of life. Kellmeyer (2019) pointed out that surgically placing implanted devices inside the human brain gives permission and high ability of these systems in some cases to share the brain thinking or processing activities and maybe making decisions [11]. That is raising a question whether a human or a machine is who would be responsible for machine errors if it happened, which leads to an unintended accident.

Additionally, the bias, justice, and normal challenges have high impact on the application of these technologies. As these devices have been manufactured and engineered by a human, to follow a specific direction of brain activity to distinguish based on for example gender, ethnicity, or color, which maybe produce interaction bias of planted brain. Consequently, these criteria could encourage more differences between society levels and its culture that places the community in a more unstable situation [13, 14]. There are different possible ways to deal with this challenge by computer and data scientists under study [15]. Also, the possible ethical concern with increasing in intelligence and memory of a specific and selected human groups with that hidden device [10]. As a result for that, the device has to be obvious, and not hidden for the society, by setting rules for that. Regarding the equality of access issues that encourage inequalities in the society, these devices have to be available to all people who need it [16–18].

4. Recommendations

There are different areas that could be used to deal with these six ethical issues, such as establishing regulations, rules, and laws depending on the public ethnicities, religious and socio-economic culture for people. Another way by enhancing and improving the technology ability, or by mixing technology and laws together depending on the conditions of a situation. For example, creating restrictive laws that prevent and minimize technologies false and exaggerated or even unsubstantiated claims, to define the lawful and justified work. Also, establishing high transparency laws of using these devices to avoid the deception with or without intention, and to make and encourage equality and fair in the society.

Additionally, developing and improving in surgery robots technologies are necessary to minimize the physical safety concerns in the inserting and establishing process of BCI devices inside the brain. Also, the scientists and suppliers community should provide safe technologies that allow access to these devices only under control of the users. Also, establishing laws that protect the personal data from being used by others, and organize data marketing and transmission that will be possible to solve the related issues with the privacy, personal information, and identity.

The agency, autonomy, and responsibility issues have become a significant challenge, which could be solved by making more restrictive laws and developing these device technologies to make a clear limit between human brain and the BCI activities. It is supposed that setting rules and developing new cheap technologies are both important factors to deal with the bias, justice, and normality challenges of these devices contain BCI. For example, new technologies that are less expensive could make these devices available to all people who need them, which is making the equality of access to these devices, to sole inequalities issues in the society. The restricted laws, which apply to force people whom are using these devices to show it in a transparent way, and it has to be obvious for other people, might solve the related issues of the bias and makes it more justice.

Acknowledgements

I would like to express all my deep thanks and respect to Federica Carta (a member of Istituto Italiano di Cultura di Lima, Sardinia, Italy) for her help and cooperation in this work.

Conflict of interest

The author declares that there are no conflict of interest.

Notes/thanks/other declarations

I would like to express all my deep thanks and respect to University of Babylon, college of engineering of materials for them help and cooperation in this work.

Appendices and nomenclature

BCI	Brain-Computer Interface
USA	United States of America
FDA	Food and Drug Administration
UK	United Kingdom
LIS	Locked-in Syndrome
USB-C	Universal Serial Bus Type-C
UNESCO	United Nations Educational, Scientific and Cultural Organization
EU	European Union
Wi-Fi	Wireless Fidelity

Author details

Akram Jassim Jawad[1,2*]

1 Department of Polymer and Petrochemicals Industrials Engineering, College of Materials Engineering, University of Babylon, Al Hillah, Iraq

2 School of Engineering and Materials Science, Queen Mary University of London, London, UK

*Address all correspondence to: akrammaterials4@gmail.com

References

[1] Fourneret E. The Hybridization of the Human with Brain Implants: The Neuralink Project. Cambridge Quarterly of Healthcare Ethics. 2020 Oct;29(4):668-672.

[2] Neuralink Progress Update, Summer 2020, Neuralink, 28 August 2020, accessed 18 December 2020.

[3] Collinger JL, Wodlinger B, Downey JE, Wang W, Tyler-Kabara EC, Weber DJ, McMorland AJ, Velliste M, Boninger ML, Schwartz AB. High-performance neuroprosthetic control by an individual with tetraplegia. The Lancet. 2013 Feb 16;381(9866):557-564.

[4] Rainey S, Maslen H, Mégevand P, Arnal L, Fourneret E, Yvert B. Neuroprosthetic speech: The ethical significance of accuracy, control and pragmatics. Cambridge Quarterly of Healthcare Ethics. 2019 Sep 2;28(S4).

[5] Kellmeyer P. Big brain data: On the responsible use of brain data from clinical and consumer-directed neurotechnological devices. Neuroethics. 2018 May 19:1-6.

[6] Steinert S, Friedrich O. Wired emotions: Ethical issues of affective brain–computer interfaces. Science and engineering ethics. 2020 Feb;26(1):351-367.

[7] Musk E. An integrated brain-machine interface platform with thousands of channels. Journal of medical Internet research. 2019;21(10):e16194.

[8] Yuste R, Goering S, Bi G, Carmena JM, Carter A, Fins JJ, Friesen P, Gallant J, Huggins JE, Illes J, Kellmeyer P. Four ethical priorities for neurotechnologies and AI. Nature News. 2017 Nov 9;551(7679):159.

[9] Coin A, Mulder M, Dubljević V. Ethical Aspects of BCI Technology: What Is the State of the Art?. Philosophies. 2020 Dec;5(4):31.

[10] Dadia T, Greenbaum D. Neuralink: The Ethical 'Rithmatic of Reading and Writing to the Brain. AJOB neuroscience. 2019 Oct 2;10(4):187-189.

[11] Kellmeyer P. 2019. Artificial intelligence in basic and clinical neuroscience: opportunities and ethical challenges. Neuroforum, 25(4), pp. 241-250.

[12] Weinberger S, Greenbaum D. Are BMI prosthetics uncontrollable Frankensteinian monsters?. Brain-Computer Interfaces. 2016 Jul 2;3(3):149-155.

[13] Knight W. Biased algorithms are everywhere, and no one seems to care. Technology Review. 2017 Jul 12.

[14] Jawad AJ. Bioethics of Medical Devices Based on Brain Computer Interfaces (BCI). J Clinic Res Bioeth. 2021; S8. 003.

[15] Courtland R. The Bias Detectives as Machine Learning Infiltrates Society, Scientists Grapple with How to Make Algorithms Fair. Nature. 2018;558(7710):357-360.

[16] Inglesant P, Hartswood M, Jirotka M. 2016. Thinking ahead to a world with quantum computers. The landscape of responsible research and innovation in quantum computing.

[17] Abdulkader SN, Atia A, Mostafa MS. Brain computer interfacing: Applications and challenges. Egyptian Informatics Journal. 2015 Jul 1;16(2):213-230.

[18] Jawad AJ. Engineering Ethics of Neuralink Brain Computer Interfaces Devices. Ann Bioethics Clin App 2021; 4(1):000160.

Chapter 9

Good Pharmacy Practice in India: Its Past, Present and Future with Need and Status in COVID 19

Mrinmoy Roy

Abstract

The pandemic of COVID-19 has highlighted the importance of emergency preparedness and response (EP and R) in India's education, training, capacity building, and infrastructure growth. Healthcare professionals, especially pharmacy professionals (PPs) in India, continued to provide drugs, supplies, and services during the pandemic. The public-private healthcare system in India is complicated and of varying quality. Patients face problems as a result of gaps in pharmacy practice education and training, as well as a lack of clarity about pharmacists' positions. Job requirements and effective placement of healthcare professionals in patient care, as well as on (EP and R) task forces or policy representation, are complicated by this lack of distinction. We have also seen malpractice and spurious distribution in the healthcare and pharmaceutical domain in terms of personal protective kits, medications, injectable, life-saving oxygen, and other items during this unprecedented pandemic situation. A few of the incidents are as follows. The central division police in Bangalore (the Global BPO & IT Hub of India) booked a case of bed-blocking at a private hospital and arrested three people, one of whom is an Arogya Mitra (primary contact for the beneficiaries at every empaneled hospital care provider), for allegedly extorting ₹1.20 lakh from the son of a COVID-19 patient who later passed away. At least 178 COVID-19 patients in India have died because of oxygen shortage in recent weeks. Another 70 deaths have been attributed to an oxygen shortage by patients' families, but this has been denied by the authorities. The Allahabad High court made a remark "Death of COVID patients due to non-supply of oxygen not less than genocide" on reports circulating on social media regarding the death of COVID-19 patients due to lack of oxygen in Lucknow and Meerut. A day ago, the Delhi police busted an industrial manufacturing unit in Uttarakhand's Kotdwar where fake Remdesivir injections were being manufactured and arrested five people. These depict the ground reality and ethical standards of good pharmacy practice in this country. There is an utmost necessity to relook and re-establish the standards of pharmacy practice in healthcare setups available in each and every corner of the country in line with guidelines provided by the World Health Organization (WHO) and the International Pharmaceutical Federation (FIP). For that, the dependency and responsibilities are very high on healthcare professionals, particularly in this pandemic situation. The pharmacy zone is

IntechOpen

adaptable, evolving, and increasingly diverse, offering a wide range of work and management opportunities to execute. PPs are human service professionals whose responsibilities include safeguarding individuals by dispensing medications based on prescriptions. Representing the world's third-largest medicinal services with active gathering, and in India, there are over 1,000,000 (1 million) enrolled PPs employed in various capacities and readily contributing to the country's well-being. Pharmacy practice, which includes clinical, community, and hospital pharmacy, is referred to as total healthcare in its true sense. Through adaptation and implementation of GPP in healthcare setup, PPs form an essential link between physicians, nurses, and patients in the social community group, with an ultimate emphasis on patient well-being and protection. To instill quality and raise the standard in this chaotic situation there are strict measures required in the country. The International Pharmaceutical Federation and World Health Organization define good pharmacy practice (GPP) as practices that meet the personal needs of patients or those using pharmacy services by offering appropriate evidence-based care. In developed countries, pharmaceutical assistance is defined as a pharmaceutical practice model that involves attitudes, ethical values, behaviors, skills, appointments, and co-responsibility to prevent diseases, promote and recovery health in an integrated manner as part of the healthcare process, highlighting, among other, the requirement that the institution fully adopts the GPP. There is a need for a GPP Program designed by the Indian Govt. or its stakeholders in the context of the Indian healthcare system and adopting "new normal" due to the unprecedented event of COVID 19 and also raising the standard and importance of GPP for the healthcare professionals in the current scenario.

Keywords: GPP in India, healthcare scams in India, scams during COVID-19, healthcare malpractices in India

1. Introduction

The COVID-19 pandemic has brought attention to the necessity of emergency preparedness and response (EP&R) in India's education, training, capacity building, and infrastructure development. During the pandemic, healthcare workers, particularly pharmacy professionals (PPs) in India, continued to deliver medication, supplies, and services. In India, the public-private healthcare system is complex and of variable quality. Gaps in pharmacy practice education and training, as well as a lack of understanding about pharmacists' roles, cause complications for patients. This lack of difference complicates job requirements and successful placement of healthcare professionals in patient care, EP&R task forces, and policy representation. During this unprecedented pandemic situation, we have also observed malpractice and bogus distribution in the healthcare and pharmaceutical arena in terms of personal protective kits, medications, injectable, life-saving oxygen, and other products. The following are a few of the incidents. The central division police in Bangalore (India's Global BPO & IT Hub) filed a complaint of bed-blocking at a private hospital and arrested three people, one of whom is an Arogya Mitra (primary contact for beneficiaries at every empaneled hospital care provider), for allegedly extorting Rs. 1.20 lakh from the son of a COVID-19 patient who later died. In the last few weeks, at least 178 COVID-19 patients in India have died due to a lack of oxygen. Another 70 deaths have been linked by patients' families to a lack of oxygen, though this has been refuted by authorities. "Death of COVID patients due to non-supply of oxygen not less than genocide," the Allahabad High Court ruled in response to reports spreading on social media regarding the death of COVID-19 patients in Lucknow and Meerut due to a shortage of oxygen [1]. Five persons were arrested after the Delhi police raided an industrial manufacturing site in Uttarakhand's Kotdwar, where duplicate Remdesivir injections were being made. This reflects the reality of Good Pharmacy Practice in this country, as well as the ethical norms that apply to it. It is critical to re-evaluate and re-establish pharmacy practice standards in healthcare settings throughout the country, in accordance with World Health Organization (WHO) and International Pharmaceutical Federation (FIP) principles. As a result, healthcare workers have a great deal of reliance and responsibility, especially in current pandemic circumstances. The pharmacy zone is adaptive, growing, and becoming more diverse, providing a comprehensive range of job and management opportunities [2]. PPs are human service professionals tasked with keeping people safe by distributing drugs according to prescriptions. In India, there are about 1,000,000 (1 million) enrolled PPs working in diverse capacities and contributing to the country's well-being, making it the world's third-largest medical service with active gathering. In its real definition, pharmacy practice, which encompasses clinical, community, and hospital pharmacy, is referred to as overall healthcare. PPs form a crucial link between physicians, nurses, and patients in the social community group through the adaptation and application of GPP in healthcare settings, with the ultimate emphasis on patient well-being and protection. In order to instill quality and elevate the standard in this chaotic scenario, the country must take stringent steps. Good pharmacy practice (GPP) is defined by the International Pharmaceutical Federation and the World Health Organization as practices that fulfill the individual requirements of patients or people who use pharmacy services by providing adequate evidence-based care. Pharmaceutical assistance is defined in developed countries as a pharmaceutical practice model that includes attitudes, ethical values, behaviors, skills, appointments, and co-responsibility to prevent diseases, promote, and recover health in an integrated manner as part of the healthcare process, highlighting, among other requirements, the institution's full adoption of the GPP. In the context of the Indian healthcare system and adopting a "new normal" due to the unprecedented event of COVID-19, there is a need for a GPP program designed by the Indian Government or its stakeholders, as well as raising the standard and importance of GPP for healthcare professionals in the current scenario.

India is a developing country in southern Asia with 29 states and seven union territories, 22 nationally recognized languages, and a population of 1.38 billion. India is a rural country with an agriculture-based economy, with rural areas accounting for over 75% of the population. In recent decades, India's average income has risen dramatically, resulting in growing urbanization, improved middle-class access to a better lifestyle, and increasing awareness of health insurance. Male and female literacy rates have climbed to 82% and 65%, respectively. According to reports, the average life expectancy is 69.9 years [3]. According to World Bank research from 2020, about 400 million Indians live on less than $1.25 a day, and 44% of children are hungry. Even though the government has taken attempts to reduce baby and maternal death rates, they remain high. Ancient Hindus acknowledged pharmacy as a complementary healing profession, and they specialized in vegetable treatments. Vedic and Brahmanic medicine were the two periods of Hindu medicine. The Vedic period or Vedic age (c.1500–c.500 BCE), and was a rudimentary period. During the Vedic period, sin was thought to be a major source of disease. Between 800 BC and 1000 AD, the Brahmanic period was a high-quality time for Hindu medical education. The works of Charaka, Susruta, and Vagbhat, which are based on ancient Vedic themes, are the three great masterpieces of Brahmanic medicine [4]. The origins of pharmacy in India can be traced back to Ayurveda, which dates to 5000 BC. Lord Brahma initially taught Ayurveda, or "life science," which was further transmitted by Charaka and Sushruta [5]. The Charaka Samhita, early work on Ayurveda, focuses on vegetable goods as well as some animal and earth goods. The classification of medications in this book is based on how they affect different bodily parts [4]. In 900 AD, Tamilnadu established a hospital to treat piles, jaundice, bleeding, and tuberculosis [5]. Tantrism was a prominent philosophical and religious movement that emerged in India after the downfall of Buddhism; it brought the art and science of the production of metallic compounds, particularly mercury and sulfur, to the fore [4]. In 1811, Scotch M. Bathgate built the first chemist shop in Calcutta, which is regarded as the birthplace of pharmacy practice in India. In December 1860, Madras Medical School began pharmacy instruction, enrolling students with only middle school education; the pharmacy degree was 2 years long. Madras Medical College (MMC), Chennai, established the diploma course in pharmacy in 1874. The Bengal Chemists and Druggists Association was founded in 1926 after the Calcutta Chemists and Druggists Association was founded in 1920. The first issue of the official *Indian Journal of Pharmacy* was published in 1939. In 1868, the British monarchy released the first Indian pharmacopeia. In 1881, a formal provision for the education and examination of compounders was established in Bengal. In 1920 and 1940, master's degrees in pharmaceutics and applied chemistry were introduced, respectively. Gorakh Prasad Srivastava was the first postgraduate pharmacy student to complete the degree in 1943. In 1945, the first steps toward standardizing pharmacy education were taken. The first annual conference of the Indian Pharmaceutical Congress Association was held in 1948. The Pharmaceutical Association was India's first pharmaceutical society, founded in 1923 and renamed The Pharmaceutical Society of India after 2 years. In 1949, a West Bengal institute introduced a Diploma in Pharmacy, and the Pharmacy Council of India (PCI) approved the diploma as the minimal prerequisite for beginning pharmacy practice in 1953. In 1954, Prof. M.L. Schroff, known as the "Father of Pharmacy in India," was elected president of the PCI.

2. Health statistics

Many issues face India's healthcare system, including the need to lower mortality rates, enhance physical infrastructure, provide health insurance, and train healthcare experts and workers. A rise in communicable diseases, lifestyle diseases, and noncommunicable diseases has been recorded. Diseases such as poliomyelitis, leprosy, and newborn tetanus will be eradicated; nevertheless, certain previously controlled infectious diseases, such as dengue fever, viral hepatitis, tuberculosis, malaria, and pneumonia, have resurfaced or developed medication resistance.

Even though Indians are now more affluent as a result of the rise of the middle class, their eating habits have shifted dramatically to unhealthy, high-sugar, high-fat diets, leading to an increase in lifestyle diseases such as hypertension, cancer, and diabetes. Furthermore, India's healthcare institutions and services would be burdened by the expanding older population.

Due to a dearth of hospital beds and skilled medical personnel such as doctors, nurses, and pharmacists, the people do not have access to their medical needs. In comparison to urban areas, healthcare services are much scarcer in rural communities. Females have an adversely skewed proportion in the healthcare workforce when compared to males. In the 11th five-year plan, total healthcare expenditure (comprising state funds, private funds, and external flows) accounted for 4.1% of GDP (GDP). The 12th five-year plan (2012–2017) intends to raise public health spending from 1.1% of GDP to 2–3% of GDP [6].

3. A look at the healthcare system as a whole

The state is responsible for the healthcare system. This system is currently managed by both public and private (for profit and nonprofit) groups. Policymaking, planning, guiding, aiding, reviewing, and coordinating the activities of various provincial health authorities are all responsibilities of the federal government, as is providing financing to implement national healthcare initiatives [7]. Allopathic hospitals, hospital beds, Indian System of Medicine and Homeopathy hospitals, subcenters, Pharmacy Health Care (PHC), Community Health Center (CHC), blood banks, Eye Bank, psychiatric hospitals, and cancer hospitals are all part of India's healthcare infrastructure [7]. The Department of Ayurveda, Yoga, and Naturopathy, as well as Unani, Siddha, and Homeopathy, provide medical and healthcare services (AYUSH). The ownership of the public sector is split between the federal and state governments, municipalities, and panchayats (local governments). Dispensaries, primary health centers, subcenters, and health posts are among the facilities. Teaching hospitals, secondary-level hospitals, first-level referral hospitals (community health centers/rural hospitals), dispensaries, primary health centers, subcenters, and health posts are among the facilities. Public facilities for specific occupational groups are also included, including organized labor (Employees State Insurance Scheme), defense, government employees (Central Government Health Scheme), railways, post and telegraph, and mining, to name a few. The private sector (profit/nonprofit) is the most prevalent, with services ranging from >1000 beds to two beds. Health-related facilities are available through the federal government healthcare program in 25 cities, with 246 allopathic dispensaries [7]. Private healthcare providers are currently treating 78% of outpatients and 60% of inpatients in India. Private healthcare providers range from world-class hospitals that promote medical tourism by providing world-class treatments to international clients and Indians who can afford it to private doctors who have minimal medical expertise or formal training on the other end of the spectrum. Furthermore, the private sector controls 80% of doctors, 26% of nurses, and 49% of hospital beds, demonstrating its power.

4. The country's hospitals

In contrast to India, the number of hospitals in other nations is not given. However, according to World Bank statistics, available at https://data.worldbank.org/indicator/SH.MED.BEDS.ZS India's accessible beds per 1000 population is lower than that of a lot of other countries. As a state of concern, public health and hospitals are largely responsible for the upkeep, providing health treatment to individuals, and maintaining hospital information. However, the following details are provided below:

i. Number of Primary Health Centers (PHCs), Community Health Centers (CHCs), Sub-District/Divisional Hospitals (SDHs), District Hospitals (DHs), and beds in India, by State/UT, as uploaded by the States/UTs on the Ministry's Health Management Information System (HMIS) portal.

ii. Number of government hospitals and beds in rural and urban areas in India, broken down by state/UT, as published in National Health Portal 2018.

iii. The Ministry of AYUSH provides information on the number of AYUSH hospitals and beds in each state/UT.

iv. Total and state-by-state number of hospitals and beds maintained by the Ministry of Defense.

v. Number of railway hospitals and beds, as reported in the National Health Portal 2018 publication.

vi. Number of State Insurance Corporation employees, hospitals, and beds, per state/UT, and total, as published in the National Health Portal 2018.

State/UT–wise number of PHCs, CHCs, SDHs, DHs, and beds in the country.

State/UT	No. of public facilities					No. of beds available in public facilities
	PHC	CHC	SDH	DH	Total	
Andaman & Nicobar Islands	27	4		3	34	1246
Andhra Pradesh	1417	198	31	20	1666	60,799
Arunachal Pradesh	122	62		15	199	2320
Assam	1007	166	14	33	1220	19,115
Bihar	2007	63	33	43	2146	17,796
Chandigarh	40	2	1	4	47	3756
Chhattisgarh	813	166	12	32	1023	14,354
Dadra & Nagar Haveli	9	2	1	1	13	568
Daman & Diu	4	2		2	8	298
Delhi	534	25	9	47	615	20,572
Goa	31	4	2	3	40	2666
Gujarat	1770	385	44	37	2236	41,129
Haryana	500	131	24	28	683	13,841
Himachal Pradesh	516	79	61	15	671	8706
Jammu & Kashmir	702	87		29	818	11,342
Jharkhand	343	179	13	23	558	7404
Karnataka	2547	207	147	42	2943	56,333
Kerala	933	229	82	53	1297	39,511
Lakshadweep	4	3	2	1	10	250
Madhya Pradesh	1420	324	72	51	1867	38,140
Maharashtra	2638	430	101	70	3239	68,998
Manipur	87	17	1	9	114	2562
Meghalaya	138	29		13	180	4585
Mizoram	65	10	3	9	87	2312
Nagaland	134	21		11	166	1944
Odisha	1360	377	27	35	1799	16,497
Puducherry	40	4	5	4	53	4462
Punjab	521	146	47	28	742	13,527
Rajasthan	2463	579	64	33	3139	51,844
Sikkim	25	2	1	4	32	1145

State/UT	No. of public facilities					No. of beds available in public facilities
	PHC	CHC	SDH	DH	Total	
Tamil Nadu	1854	385	310	32	2581	72,616
Telangana	788	82	47	15	932	17,358
Tripura	114	22	12	9	157	4895
Uttar Pradesh	3277	671		174	4122	58,310
Uttarakhand	275	69	19	20	383	6660
West Bengal	1374	406	70	55	1905	51,163
All India	**29,899**	**5568**	**1255**	**1003**	**37,725**	**739,024**

Source: Data as uploaded by States-UTs on HMIS portal, status as of 20 July 2018.

India, State/UT wise number of government hospitals and beds in rural and urban areas.

States/UTs	Rural hospitals		Urban hospitals		As on
	No.	Beds	No.	Beds	
Andhra Pradesh	193	6480	65	16,658	01.01.2017
Arunachal Pradesh*	208	2136	10	268	31.12.2017
Assam	1176	10,944	50	6198	31.12.2017
Bihar	930	6083	103	5936	31.12.2016
Chhattisgarh	169	5070	45	4342	01.01.2016
Goa*	17	1405	25	1608	31.12.2017
Gujarat	364	11,715	122	20,565	31.12.2016
Haryana*	609	6690	59	4550	31.12.2016
Himachal Pradesh*	705	5665	96	6734	31.12.2017
Jammu & Kashmir	56	7234	76	4417	30.12.2016
Jharkhand	519	5842	36	4942	31.12.2015
Karnataka*	2471	21,072	374	49,093	31.12.2017
Kerala	981	16,865	299	21,139	01.01.2017
Madhya Pradesh	334	10,020	117	18,819	01.01.2016
Maharashtra	273	12,398	438	39,048	31.12.2015
Manipur	23	730	7	697	01.01.2014
Meghalaya*	143	1970	14	2487	31.12.2017
Mizoram*	56	604	34	1393	31.12.2017
Nagaland	21	630	15	1250	31.12.2015
Odisha*	1655	6339	149	12,180	31.12.2017
Punjab*	510	5805	172	12,128	31.12.2017
Rajasthan	602	21,088	150	10,760	31.12.2016
Sikkim*	24	260	9	1300	31.12.2017
Tamil Nadu*	692	40,179	525	37,353	31.12.2017
Telangana*	802	7668	61	13,315	31.12.2017

States/UTs	Rural hospitals		Urban hospitals		As on
	No.	Beds	No.	Beds	
Tripura*	99	1140	56	3277	31.12.2017
Uttar Pradesh*	4442	39,104	193	37,156	31.12.2017
Uttarakhand	410	3284	50	5228	31.12.2015
West Bengal	1272	19,684	294	58,882	01.01.2015
Andaman & Nicobar Islands	27	575	3	500	31.12.2016
Chandigarh	0	0	4	778	31.12.2016
Dadra & Nagar Haveli*	10	273	1	316	31.12.2017
Daman & Diu	5	240	0	0	31.12.2015
Delhi	0	0	109	24,383	01.01.2015
Lakshadweep	9	300	0	0	01.01.2016
Puducherry	3	96	11	3473	01.01.2016
INDIA	19,810	279,588	3772	431,173	

Source: National Health Profile 2018/Directorate General of State Health Services.
Notes: Government hospitals include central government, state government, and local govt. bodies. Figures are provisional.
**States/UTs provided information for the year 2017 and PHCs are also included in the number of hospitals.*

State-wise distribution of AYUSH hospitals and beds as on 1-4-2017.

Srl no.	State/UT	Number of hospitals				Number of beds			
		Govt.	Local Body	Others	Total	Govt.	Local Body	Others	Total
1	Andhra Pradesh	8	0	0	8	365	0	0	365
2	Arunachal Pradesh	11	0	1	12	100	0	25	125
3	Assam	4	0	0	4	205	0	0	205
4	Bihar	8	0	0	8	950	0	0	950
5	Chhattisgarh	7	0	8	15	370	0	470	840
6	Delhi	4	0	1	5	390	0	210	600
7	Goa	0	2	0	2	0	150	0	150
8	Gujarat	35	29	0	64	1495	1763	0	3258
9	Haryana	4	0	7	11	145	0	600	745
10	Himachal Pradesh	34	0	0	34	710	0	0	710
11	Jammu & Kashmir	2	0	0	2	75	0	0	75
12	Jharkhand	1	4	0	5	50	300	0	350
13	Karnataka	156	0	57	213	2455	0	7740	10,195
14	Kerala	162	0	0	162	4054	0	0	4054
15	Madhya Pradesh	23	0	0	23	690	0	0	690
16	Maharashtra	5	0	66	71	812	0	8024	8836
17	Manipur	10	0	0	10	0	0	0	0
18	Meghalaya	10	0	0	10	100	0	0	100

Srl no.	State/UT	Number of hospitals				Number of beds			
		Govt.	Local Body	Others	Total	Govt.	Local Body	Others	Total
19	Mizoram	2	0	0	2	100	0	0	100
20	Nagaland	2	0	0	2	10	0	0	10
21	Odisha	9	0	5	14	593	0	310	903
22	Punjab	5	0	4	9	100	0	108	208
23	Rajasthan	137	0	0	137	1391	0	0	1391
24	Sikkim	1	0	0	1	10	0	0	10
25	Tamil Nadu	293	0	0	293	2381	0	0	2381
26	Tripura	4	0	0	4	60	0	0	60
27	Uttar Pradesh	2315	0	1	2316	11,361	0	100	11,461
28	Uttarakhand	404	0	1	405	2023	0	50	2073
29	West Bengal	11	0	11	22	530	0	620	1150
30	Andaman & Nicobar Islands	5	0	0	5	90	0	0	90
31	Chandigarh	0	0	2	2	0	0	80	80
32	Dadra &Nagar Haveli	0	0	0	0	0	0	0	0
33	Daman &Diu	2	0	0	2	20	0	0	20
34	Lakshdweep	0	0	0	0	0	0	0	0
35	Puducherry	0	0	0	0	0	0	0	0
36	Telangana	20	0	0	20	841	0	0	841
	Total (A)	3694	35	164	3893	32,476	2213	18,337	53,026
B. CGHS & central government organizations		50	0	0	50	2216	0	0	2216
	Total (A+B)	3744	35	164	3943	34,692	2213	18,337	55,242

Source: Ministry of AYUSH.

Number of hospitals and beds maintained by Ministry of Defense, State wise.

S. no.	Name of state	No. of hospitals	No. of beds
1	Assam	8	2357
2	Andhra Pradesh	1	306
3	Andaman & Nicobar Islands	1	107
4	Arunachal Pradesh	1	198
5	Bihar	2	348
6	Delhi	2	1993
7	Goa	2	175
8	Gujarat	5	666
9	Haryana	3	1458
10	Himachal Pradesh	6	699

S. no.	Name of state	No. of hospitals	No. of beds
11	Jammu & Kashmir	11	2643
12	Jharkhand	2	649
13	Karnataka	3	1090
14	Kerala	5	744
15	Madhya Pradesh	7	1402
16	Maharastra	11	4202
17	Manipur	1	74
18	Meghalaya	1	247
19	Nagaland	2	398
20	Odisha	2	147
21	Punjab	10	2990
22	Rajasthan	11	2115
23	Sikkim	2	247
24	Tamil Nadu	3	373
25	Telangana	3	764
26	Tripura	1	49
27	Uttarakhand	5	1402
28	Uttar Pradesh	15	4570
29	West Bengal	7	2107
	Total	133	34,520

Source: Ministry of Defense

Number of hospitals and beds in railways (as on 21/03/2018).

S. no.	Zone/PU	Total no. of hospitals	Total no. of indoor beds
1	Central Railway	11	1164
2	Eastern Railway	8	1587
3	East central Railway	9	819
4	East coast Railway	4	339
5	Northern Railway	9	1101
6	North Central Railway	5	586
7	North East Railway	6	927
8	North East Frontier Railway	10	1107
9	North Western Railway	8	584
10	Southern Railway	10	1131
11	South Central Railway	7	714
12	South Eastern Railway	6	1086
13	South East Central Railway	5	250
14	South Western Railway	3	300
15	Western Railway	9	976

S. no.	Zone/PU	Total no. of hospitals	Total no. of indoor beds
16	West Central Railway	7	456
17	Intergral Coach Factory	1	101
18	Rail Coach Factory	1	60
19	Chittaranjan Locomotive Works	1	197
20	Diesal Locomotive Works	1	105
21	Diesel Loco Modernisation Works	1	50
22	Rail Wheel Factory	1	46
23	Research Design and Standards Organization	1	30
24	Metro/Kolkata	1	30
25	MCF/Raibareli	1	2
26	**Total**	**126**	**13,748**

Source: National Health Profile 2018 (as on 21/3/2018).

Employees State Insurance Corporation hospitals and beds (as on 31.03.2017).

S. no.	States/UTs	Total no. of hospital	Total no. of beds
1	Andhra Pradesh	5	345
2	Assam	1	75
3	Bihar	3	50
4	Chandigarh [Adm.]	1	70
5	Chhattisgarh	0	0
6	Delhi	4	1416
7	Goa	1	15
8	Gujarat	12	910
9	Himachal Pradesh	2	150
10	Haryana	7	781
11	Jammu & Kashmir	1	50
12	Jharkhand	3	210
13	Karnataka	11	1675
14	Kerala	12	1178
15	Madhya Pradesh	7	725
16	Meghalaya	0	0
17	Maharashtra	13	2390
18	Nagaland	0	0
19	Odisha	6	325
20	Puducherry	1	75
21	Punjab	8	647
22	Rajasthan	6	495
23	Sikkim	0	0

S. no.	States/UTs	Total no. of hospital	Total no. of beds
24	Tamil Nadu	10	1856
25	Telangana	7	907
26	Tripura	0	0
27	Uttar Pradesh	16	1886
28	Uttarakhand	0	0
29	West Bengal	14	3534
	Total	**151**	**19,765**

Source: National Health Profile 2018.

5. Core pharmacy practices

5.1 Hospital pharmacy

The hospital pharmacy is one of the most important departments in the hospital, and it is responsible for drug procurement, storage, compounding, dispensing, manufacturing, testing, packaging, and distribution. This section is also in charge of pharmaceutical science and education research, which is carried out by skilled and knowledgeable pharmacists. The hospital pharmacy has a significant impact on healthcare cost economics. In today's hospital pharmacy, medication monitoring and drug information services are combined. Purchasing, storing, handling, pricing, and dispensing pharmaceuticals are all skills that a hospital pharmacist possesses. In addition, pharmacists give drug information to all healthcare professionals and the public, as well as serve as a link between the patient and the doctor. The criteria for acquiring pharmaceuticals, chemical and biological medications, and other items are provided by hospital pharmacists. They are also in charge of manufacturing and distributing pharmaceuticals like transfusion fluids, parenteral goods, pills, capsules, ointments, stock combinations, and safe drug storage. Hospital pharmacists can sterilize and dispense parenteral drugs made in hospitals. They fill, label, and distribute all medicine packages. Hospital pharmacists oversee purchasing pharmaceuticals, ensuring correct drug storage conditions, keeping records, and distributing pharmaceuticals to the outpatient department. In addition, hospital pharmacists provide drug monitoring services for inpatients and participate in hospital research programs [8]. Due to low wages, the hospital parts of a pharmacist's job have historically been overlooked in India; also, pharmacists have never been trained for a patient-centered role. Medical practitioners have never embraced pharmacists for clinical functions, while pharmacists have been hesitant to accept their profession's clinical obligations. Many Indian hospitals have begun to assign pharmacists a clinical role, with encouraging outcomes, but India still lags behind other countries in this regard. In India, the concept of a hospital pharmacy is confined to the dispensing of medications at hospital pharmacies [9].

Hospital pharmacies require certain personnel.

The dispensing, manufacturing, quality assurance, and clinical pharmacy services are all integrated within the hospital pharmacy. Personnel requirements for an inpatient pharmacy are determined by the nature and scope of services supplied by the department. The number of hospital pharmacists required is determined by the workload and the number of beds in the hospital. Small hospitals typically require a minimum of three pharmacists; however, this varies depending on the number of beds. **Table 1** shows the number of pharmacists necessary based on the number of beds in a hospital [8].

According to 2011 data, the number of government hospital beds in rural and urban locations is shown in **Table 2**.

Personnel who work in hospital pharmacies must be knowledgeable and well-trained. The director of the pharmacy must have a postgraduate degree in hospital pharmacy, pharmacology, or pharmaceutics and serves as a liaison between pharmacy and non-pharmacy personnel. The hospital pharmacy's structure is depicted in the Flowchart below (**Figure 1**).

Even we have seen the present status, there were no significant improvements.

5.2 Community pharmacy

In India, the Pharmacy Act of 1948 was passed to ensure that every practicing pharmacist had a registration certificate [11–13]. On completion of the minimum diploma in pharmacy, educational institutes authorized by the PCI can grant this

Pharmacist requirement	
Bed strength	**No. of pharmacists required**
Up to 50 beds	3
Up to 100 beds	5
Up to 200 beds	8
Up to 300 beds	10
Up to 500 beds	15

Table 1.
Pharmacist requirement in hospitals.

State	Rural hospital beds (government)	Urban hospital beds (government)	Total beds (government)	Proportion of rural and urban beds
Bihar	1830	16,686	18,516	10:90
Chhattisgarh	3270	6158	9428	35:65
Jharkhand	N.A.	N.A.	N.A.	N.A.
Madhya Pradesh	10,040	18,493	28,533	35:65
Odisha	7099	8715	15,814	45:55
Rajasthan	13,754	12,236	25,990	53:47
Uttar Pradesh	15,450	40,934	56,384	27:73
Uttarakhand	3746	4219	7965	47:53
EAG states	55,189	107,477	162,630	34:66
Non-EAG states	114,673	511,187	622,310	18:82
All India	169,862	618,664	784,940	20.5:79.5

N.A., not available.
Source: GOI, Table 6.2.2 State/UT wise number of govt. hospitals and beds in rural and urban areas (including CHCs) in India (provisional), in "Health infrastructure" in "National Health Profile, 2011," Central Bureau of Health Intelligence, Ministry of Health and Family Welfare, 2011.

Table 2.
State and union territory wise number hospitals and beds in rural and urban areas in India.

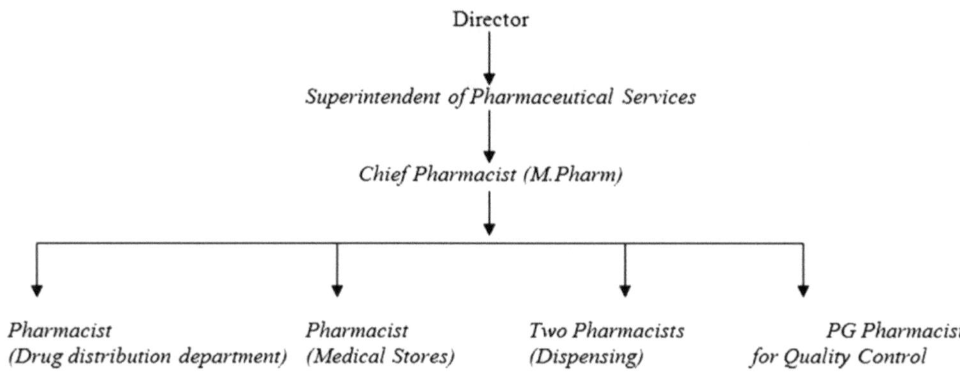

Figure 1.
Organization structure of hospital pharmacy. Source: Nahata [10].

credential. A pharmacy diploma involves at least 2 years of education and 500 h of practical training, including three months in a community or hospital pharmacy [11]. These pharmacists with a diploma represent most of the workforce and run most community pharmacies in India today. A bachelor's degree in pharmacy is designed to meet the needs of the pharmaceutical business, with most graduates working in pharmaceutical companies or regulatory agencies. In comparison to community pharmacy, many B. Pharm graduates prefer to work in the pharmaceutical business because of the higher pay scale and other benefits.

In practice, these pharmacists with a diploma or bachelor's degree are still unheard of in half of the community pharmacies. Most patients come to community pharmacists for help on sexually transmitted diseases, minor ailments, contraceptive options, and menstruation issues [11]. The majority of community pharmacies are administered by people with minimal awareness of health issues or medical training, and pharmacists are hired on low pay rates. Most pharmacists in community pharmacies lack counseling abilities and are simply able to deliver medications. Despite the public's negative opinion of pharmacists and the idea that pharmacies are like grocery stores, community pharmacies remain the principal provider of low-cost medical treatment.

5.3 Medicine promotion and marketing

In India nowadays, the average cost of a prescription per patient who visited a private clinic is around 1000–1500 rupees, and the number of prescriptions per drug has increased. The marketing efforts of pharmaceutical companies have a significant impact on doctors' prescribing patterns. Pharmaceutical companies spend less on R&D than they do on marketing. Medical representatives are trained by pharmaceutical companies to advertise and sell pharmaceuticals utilizing printed product materials, medicine samples, and gifts. Around 80,000 medical representatives are hired by pharmaceutical companies in India, and they are paid well.

In addition to their pay, they are offered significant incentives. General practitioners have been found to prescribe new products under the influence of sales reps. Furthermore, medical representatives promote medications by stating more indications than are officially registered.

One of the main tools used by pharmaceutical firms to advertise their medication is gifts (ranging from a table-top reminder to an air conditioner), gratification, foreign excursions, pleasure excursions, and so on. From clocks to air conditioners, calendars to cars, rubber bands to refrigerators, telephone indexes to television, and office items to overseas trips, the variety of gifts includes stationery, timepieces, bags, books, folders, office desks, medical supplies, and household items—from clocks to air conditioners, calendars to cars, rubber bands to refrigerators, telephone indexes to television, and office items to overseas trips. According to Chren and associates, there is an implicit relationship between doctors and the pharmaceutical industry, which leads to a responsibility to respond to gifts [14].

Advertising has a significant role in the marketing of medicine. Pharmaceutical companies spend a lot of money to advertise in medical and pharmaceutical magazines, which are read by a lot of general practitioners and clinical specialists. Medical journals in India would perish if pharmaceutical companies did not advertise in them. The advertising of pharmaceuticals is strictly monitored, and if there is any false information, harsh action is taken. Drug information should be accurate, current, and balanced, according to the International Federation of Pharmaceutical Manufacturers Associations (IFPMA) and The Association of the British Pharmaceutical Industry (ABPI) guidelines. The IFPMA code must be followed by all members of the Organization of Pharmaceutical Producers of India (OPPI). Internal codes and promotional material for the pharmaceutical producers of India must be approved by medical advisers, according to the organization. A Promotional Quality Improvement and Assurance Committee also performs a thorough study of these promotional materials. Furthermore, this committee makes recommendations for marketing enhancements.

Irrational prescription and its consequences are heavily influenced by pharmaceutical marketing and promotional activities. The Medical Council of India has issued a new code of ethics for practitioners who receive any sort of compensation from pharmaceutical companies. New legislation in Rajasthan, India, states that a doctor must only put the generic name of a medicine on a prescription [14].

5.4 Pharmacy-related services and activities that are unique

With the advent of clinical pharmacy, which provides high-quality treatment and assistance to patients, a pharmacist's job has been expanded. Pharmacists now provide drug and toxicity information, manage anticoagulation clinics, and administer smoking cessation programs, among other things.

6. Education parameters of healthcare workforce

The paramedic education system has been supply-driven, with no regard for industry demand. Allied healthcare providers are the most crucial because, if a patient has a hundred interactions at a hospital, allied healthcare workers account for 80 of those encounters. In India, healthcare is the third most important source of employment. It is broken down into three categories: Doctors make up 10–15% of the healthcare workforce, nurses make up 20–25% of the workforce, and paramedics, also known as allied healthcare professionals or healthcare technicians, make up the largest segment of the healthcare workforce. The doctor-to-allied healthcare-professional ratio in India is 1:4. The advanced countries, such as the United States and the United Kingdom, have a 1:20 ratio. India already employs over 10 million allied healthcare workers, but to meet the 1:20 ratio, we will need to increase this number to 40–50 million in the next few years. Diagnostic, curative, and rehabilitative allied healthcare professions are categorized into three groups. The diagnostic and therapeutic industries in India have been expanding. Allied healthcare professionals such as X-ray technicians and medical app technicians work in diagnostic. Technicians such as OT technicians and emergency medical technicians work in Curative. The difficulty is that in India, only about 7% of these allied healthcare professionals are certified because there was no centralized authority to certify them until the NSTC was established. Before NSTC, each state certified them at their own level, and the biggest problem today is for quality healthcare services, allied healthcare professionals are the most important because if a patient has a hundred interactions in a hospital, 80 of those interactions are with allied healthcare professionals, and the biggest challenge in India is that there are less than a hundred allied healthcare professionals. Until 2008, there was no centralized body. One of the most significant issues is that, in general, education in India has been supply-driven, with no connection to industry needs. As a result, when compared to worldwide norms, we are one of the least prepared for the workforce. What needs to be done is for education to be driven by demand. It must be presented from the standpoint of the company and how these individuals will work in the sector. To do so, we must recognize three factors—technical training, practical training, and soft skills and grooming training, all of which must be provided so that these individuals are prepared to fully contribute to the workforce in a productive manner when they enter the workforce. Mr. Narendra Modi, Prime Minister of India, believes that ongoing training of the workforce required to combat the COVID 19 crisis is critical. He recommended officials and doctors to hold training sessions and webinars, particularly for paramedics and doctors who work in rural areas. He also requested that the officials endeavor to reduce vaccination waste in the district. Interacting with Varanasi doctors and administrators via video chat Prime Minister Narendra Modi paid a heartfelt tribute to those who died because of COVID-19. He emphasized the necessity of ongoing training for those fighting COVID, advising officials and doctors to hold training sessions and webinars, particularly for paramedics and doctors functioning in remote areas. He also requested that the officials endeavor to reduce vaccination waste in the district. "Jahan Bimar Wahan Upchar," is the new slogan in COVID management in terms of delivering therapy to the patient's doorstep that will minimize the strain on the health system. The doctors from Varanasi updated the other doctors through video conference on the measures made in the last month to restrict the spread of COVID, vaccination status, and ongoing steps and strategies to prepare the district for future difficulties. They stated that they have been vigilant in their response to the threat of mucormycosis and have already taken action and established facilities to handle the condition. Our Prime Minister also praised the microcontainment zones project and praised the home distribution of medicines, and he urged health professionals to make the campaign as widespread as possible in rural areas. He praised the physicians, nurses, technicians, ward boys, ambulance drivers, and other Kashi frontline health professionals for their efforts. He praised the medical staff in Kashi for their efforts in controlling the pandemic to a large extent, but warned against complacency, urging them to fight a lengthy battle right now by focusing on the rural districts of Banaras and Purvanchal. PM emphasized the critical role of Accredited Social Health Activists (ASHAs) and Auxiliary Nursing Midwifery (ANMs) sisters in the ongoing village campaign against COVID-19 and urged health officials to use their potential and experience. The main reason is if the frontline employees had already been vaccinated, they were able to serve the people safely during the second wave. We need to encourage everyone to get the vaccine when it is their turn, that is what has currently been done.

7. International standards of good pharmacy practice

Diverse countries and continents, including emerging, transitional, and industrialized countries, have different pharmaceutical practices. The World Health

Organization (WHO)/International Pharmaceutical Federation (FIP) joint guideline of good pharmacy practice (GPP), which was updated in 2011, is meant to account for these changes in practice. The pharmacy profession is rapidly expanding, with new responsibilities being offered and announced not only by pharmacists but also by other medical professionals, national organizations, and institutions. The GPP's guideline is progressive and adaptable, and it must remain relevant when new positions develop. Standardizing many parts of pharmacies is one of the most important techniques of quality control for medications and pharmaceutical services for the public. The GPP Guide is a significant step toward bettering pharmacy services [12]. The GPP guideline is designed for use by national professional pharmaceutical associations, as well as national authorities and other relevant agencies responsible for drafting relevant documents and related laws and regulations in their countries, according to the WHO and FIP. It is not a clear national standard, but it does provide guidance on certain tasks, responsibilities, and actions that help pharmacists to fulfill their mission. Professional issues and attitudes are given specific attention throughout the guideline's content, and the patient's well-being is given priority. However, it should be noted that this is the first time that a legal, economic, and labor framework has been introduced in the context of the GPP structure, and it comes at an appropriate time given global debates on the economic aspects of medicines, access to quality medicines, access to skilled medical workers, the global workforce failure, the increased cost of medical care, and new models of healthcare [13]. The GPP is defined by the WHO and the FIP as a pharmacy practice that satisfies the needs of people who use pharmacists' services to deliver optimal medical care based on evidence-based medicine principles. To support this approach, a national framework of quality standards and guidelines is required. The goal of this project is to investigate the historical stages of development of GPP requirements, as well as to examine modern normative texts proposed by FIP for use in global pharmaceutical practice, as well as to investigate the current state of this issue around the world. The study included the generalization of information material and system analysis methods. The WHO and the FIP have been working together since the late 1980s to define pharmacist duties and functions, as well as produce guidelines for the GPP as a framework for pharmacological care. The WHO amended the medicines strategy, which was accepted by the World Health Assembly in 1986, and two meetings on the function of pharmacists were held in accordance with it. The inaugural meeting was held in Delhi, India, in 1988 [15]. The FIP produced criteria for pharmaceutical services in 1992, and the WHO Information Centers disseminated the approach of good pharmaceutical practice in March 1993. The standards of pharmacy services entitled "GPP in Public and Hospital Pharmacies" were officially issued during the convention in Tokyo, Japan [15]. Under the framework of the Tokyo declaration on quality standards of pharmaceutical services, the FIP congress accepted the FIP/GPP text. The FIP has asked pharmaceutical companies and governments to work together to implement GPP standards or to amend current national standards in countries where they currently exist, according to this paper. Then, in May 1994, the World Health Assembly adopted a resolution on the role of the pharmacist in support of the WHO's new pharmaceuticals strategy. In 1994, a version of the GPP was also presented to the WHO Expert Committee on Pharmaceutical Medicine Specifications in Geneva. In 1999, a joint document of the WHO/FIP of the GPP was published in a series of technical reports of the WHO Expert Committee on concerns of pharmaceutical medicine specification, following the recommendations of the WHO Expert Committee and the approval of the Board of the FIP in 1997. This gave the GPP more formal standing and assured that it was widely distributed around the world [16]. A Soon after, the WHO convened two more meetings on the pharmacist's role, one in 1997 in Vancouver, Canada, and the other in 1998 in The Hague, Netherlands. These gatherings reinforced the necessity for reforming pharmacy education programs and highlighted the pharmacist's role in self-help and self-healing. The European Union's (EU) pharmaceutical group prepared a paper containing the GPP for Europe in 1998, with special emphasis dedicated to EU countries—"the GPP in Europe." [17]. The "GPP in the New Independent States" was established in 2001. The Copenhagen Center for Drug Policy and Development of Pharmaceutical Practice developed "Guidelines for the creation and implementation of standards" in Denmark. The document examined the state of pharmacy practice and focused on issues such as health education and morbidity prevention, the provision of prescription drugs and their use, self-treatment, the effects of prescription and use of drugs, as well as the method of implementing proper pharmacy practices in developing countries. The WHO experts recommended that European standards for pharmacies be implemented on a par with existing national standards to regulate various aspects of pharmaceutical activity, including the quality of prescription data received by the pharmacist, the development of medical forms, the building of contacts with doctors based on individual recipes, and the evaluation of data on the use of medical products [18]. There is a strategy for establishing national GPP standards, according to FIP recommendations. To begin, fundamental pharmaceutical services must be established, for which a regulatory framework and relevant standards must be created. The secondary and higher education systems in pharmacy should be revised after that because pharmacists will require a more in-depth understanding of pharmacotherapy, pharmacopeia, and communications skills. The provision of more complex professional pharmaceutical services is the ultimate stage. The FIP highlights GPP elements that demonstrate that the pharmacy provides high-quality service and adheres to strict guidelines: The presence of a zone in the pharmacy with information about healthy lifestyles; availability of a pharmacy for people with disabilities and elderly patients; availability of a comfortable waiting area; the possibility of a private conversation between a pharmacist and a patient, including those with disabilities; and the presence of a pharmacist on call. The pharmacist's responsibility in providing successful drug therapy, according to FIP's recommendations, is as follows: Management of therapy, monitoring of treatment success, and disseminating information on the safe and effective use of medications. The pharmacist should evaluate the client's health and needs, taking into consideration his unique qualities. Professional collaboration between the pharmacist and the doctor is an important part of GPP. Each patient's medical and pharmacological information (diagnosis and laboratory data) should be in the first one. It is also crucial to keep track of whether GPP standards are being followed. It can be internal and/or external; it can also be necessary or simply for accreditation/certification purposes. This will disclose the drawbacks of meeting GPP requirements as well as ideas for how to avoid them. One of the simplest ways to ensure that a pharmacy's work is of high quality is to assess customer satisfaction with the service. In 2006, FIP and WHO collaborated to publish the handbook "Developing Pharmacy Practice—A Patient-Centered Approach." This guide introduces the new pharmacy practice paradigm and the approach to pharmaceutical assistance. The FIP has taken the initiative to explore the possibilities of providing technical assistance to its member organizations in Cambodia, Moldova, Mongolia, Paraguay, Thailand, Uruguay, and Vietnam in developing national standards of the GPP in a pilot project to improve standards and practice of medicine distribution and use with the help of the WHO/FIP setting with the GPP as the framework. The "Bangkok declaration of the GPP in public pharmacies" was adopted across Southeast Asia in 2007, and member associations pledged to improve the quality of pharmacy services and professional practice [19].

Significant advances in practice applied science and technology, and pharmaceutical policy has happened since the GPP guideline was accepted in community and hospital settings. Despite changes that have occurred since the GPP's previous guidance in pharmaceutical policy, practice, and applied science was adopted in 2007, FIP was established to investigate the problem of updating the GPP's guiding principles to reflect current standards and professional thinking peculiarities [20]. During the 68th World Congress in Basel, Switzerland, in 2008, the FIP convened expert consultations for this aim. A total of 50 people attended the meeting, including members of the working group (WG), FIP of the GPP, WHO headquarters staff, representatives from the WHO regional office for the Eastern Mediterranean countries, medicines advisers from Ghana, Nigeria, and the United Republic of Tanzania, presidents, and secretaries of six Regional Pharmaceutical Forums, FIP member organizations, and several NGOs [21]. Following these consultations, the FIP GPP working group performed a detailed examination of current national standards for GPP in at least 37 countries and developed a time zone that could allow for adequate consultations with all 120 national associations, relevant experts, and the WHO. In October 2008, a proposal for this project was given to the WHO Expert Committee on concerns of pharmaceutical medicine specification at its 43rd meeting, and an updated report was provided to the Committee of Experts at its 44th meeting in October 2009. Simultaneously, in late 2008, the Pan-American Health Organization prepared “the guide for pharmaceutical services in primary health care” with the help of a group of experts from various pharmaceutical organizations, with the goal of highlighting the role of pharmaceutical experts in Latin America's health system. Following discussion with 120 national members of the FIP in 2011 and changes in the pharmaceutical industry, the GPP recommendations were adopted and updated by the approval of a joint guideline FIP/WHO titled "the GPP: Standards of the quality of pharmacy services." This general guideline was released as new standards of pharmacy service quality in the WHO Expert Committee's 45th report. In addition, this document encourages national pharmacy professional associations to embrace these standards and propose some GPP-specific norms [22]. The GPP establishes standards that are usually higher than the requirements of a country's pharmaceutical legislation. In the modified version of the GPP, pharmacists play many roles:

1. Medical product manufacturing, receiving, storage, security, distribution, usage, release, and disposal.

2. Ensuring that the pharmacological therapy is properly managed.

3. Continuation and enhancement of professional activity.

4. Encouraging the system of medical care and health to become more effective [20].

Depending on the duties that each pharmacist does, these positions may differ. Only through national pharmacy professional associations may specific GPP standards be defined. This recommendation is made in the form of a set of professional goals that must consider the needs of patients and other pharmaceutical stakeholders.

In comparison to the previously authorized concept of the GPP, the present updated version reinforces the requirements for the GPP's well-known primary aspects and identifies functions in each pharmacist's position for which minimum national standards should be established. Different countries have different versions of the GPP rules. There are holistic papers in some countries that cover both requirements for the material-technical base, infrastructure, personnel, and standards for the provision of pharmacological treatment. These rules and regulations are spelled out in numerous documents in other countries. The standards for pharmacists in France, for example, are established in several guidelines. In Austria, on the other hand, nearly all the requirements for pharmacists are encapsulated in a single piece of legislation. There is also a guideline for pharmacists' work, which includes standards for a quality management system in addition to the typical criteria for premises, equipment, and staff. The GPP standards have a recommended status in industrialized countries. As a result, the Norwegian pharmaceutical association's standards of pharmacy practice, created in partnership with other professional groups, include requirements for pharmacy activities that are utilized by pharmacy owners to conduct internal quality control of pharmacy services. The government adopts legislation acts that provide the minimum standards for the functioning of pharmacies [23]. The Pharmaceutical Society of Ireland published a guideline on pharmacy practice to help pharmacists meet legal and regulatory requirements when providing pharmaceutical services [24]. In addition, the GPP is adopted in CIS countries. Belarus has implemented a pharmacy classification system. The GPP was adopted in Kazakhstan in 2006, and the provisions of the GPP of the customs union are currently approved in the Russian Federation. It should be noted that the GPP standard and guidance on the implementation of proper pharmacy practices serve as the foundation for implementing the total quality management (TQM) concept and the international standard of ISO quality management at pharmaceutical companies, both of which have been in use for more than 30 years. The understanding of the existence of an inextricable link between the quality management system and the organization's management system, as well as the understanding that this is an essential tool for continuous improvement and increasing the pharmacy organization's competitiveness in any market, is the foundation of TQM [25]. As a result, by enhancing the requirements to assure the quality of public services, the GPP standards are a significant step toward the expansion and enhancement of pharmacy operations. Despite improvements in pharmaceutical research and practice, the development and implementation of GPP criteria into pharmacy practice is a long-term and ongoing process. The standardization of pharmacists' activity in providing people with medications and medical items receives a lot of attention in GPP guides of all years. Because national regulation of pharmacy practice differs widely among nations, the establishment of GPP standards should be handled at the level of public professional organizations. The results of the study will be used to design and apply GPP national standards around the world, according to future scientific research.

8. Current pharmacy practice scenario in India

A recent countrywide survey done by final year postgraduate diploma in management students from Indian Institute of Health Management & Research, Bangalore finds that out of 107 hospitals 66 hospitals only follow the WHO/FIP GPP 2011 or Indian Pharmacy Practice Regulations 2015 which is merely 62% of the entire sample.

Students were asked to conduct a brief analysis and report on the top five hospitals or community pharmacy dept. in their area or where they did train/observer ship whether these guidelines are followed or not and where are the loopholes in following this guideline. There were two kinds of data generated, such as

1. **Quantitative** data by asking close-ended questions, and

2. **Qualitative** data by observation and asking open-ended questions.

The graphical representation of the quantitative survey data is illustrated as follows:

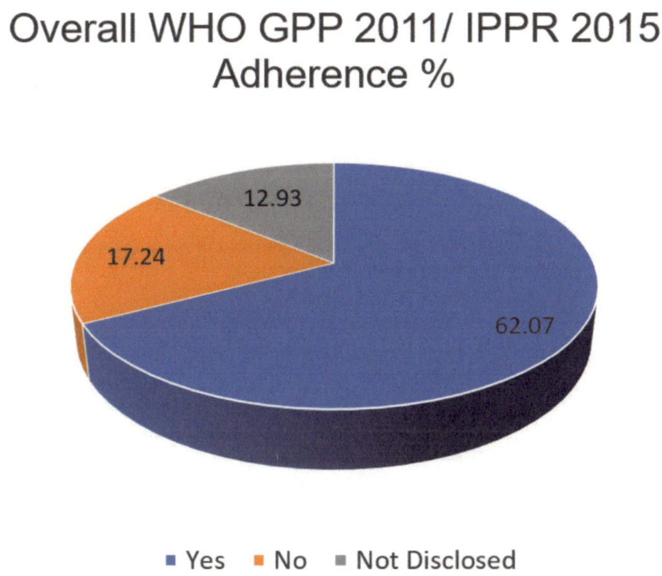

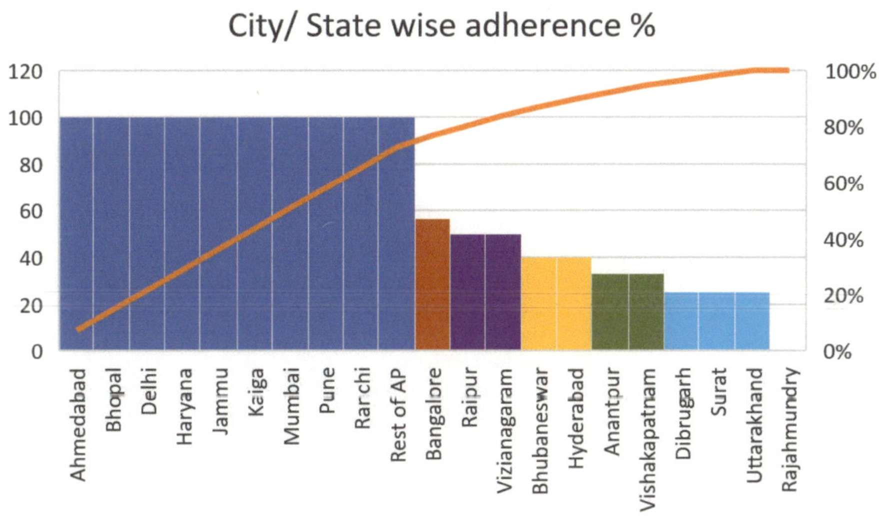

The qualitative observational/survey data is represented as follows:

Qualitative survey result	Observations
Narayana Health City	While working in a Pharmacy Department of the Hospital, many things are founded like: • Medicines are not properly arranged. • Medicines of different names and labels are found in different boxes. • Some medicines are found unsealed that is a loss of Hospital pharmacy. • One medicine was labeled in two boxes, for example, Flexurra D • There is a lack of space. • Staff are not attending calls properly. • Medicines were lying on the floor. • Pharmacy is not hygienic.

Qualitative survey result	Observations
	• There should be the proper categorization of injections and other items. This difference is due to: • Lack of coordination between the staff. • Lack of staff present in the pharmacy. • Improper arrangements of the drugs cause difficulty in searching medical items. Challenges- • Lack of commitment and coordination toward the service delivery • Lack of training, shortage of manpower and facilities for pharmacist Recommendation There should be a proper space in the IP pharmacy. Pharmacists should be provided with more space with more no of counters. Expired medicine should be categorized separately. Medicine should be categorized on a monthly basis so that it may return on time. Calls should be attended properly by the staff. The pharmacy should categorize the drugs, injections, medical consumables in an order so that anyone can find it quickly. Drinking water facilities should be there. There should be proper lighting facilities. The pharmacy should be hygienic.
Dr. Agarwal's Eye Hospital Chain, S V Hospital & C M N Hospital	a. There is no particular pharmacist assigned to the pharmacy. b. The pharmacy is left open and can be accessed by anyone including patients. c. There is no proper maintenance of the record of the drugs in the pharmacy. d. The hospital is not departmentalized, hence there is no specific manager for the pharmacy. e. The drug availability is poor and sometimes they borrow from the nearest medical store. The prescribed medicines are not always available in the pharmacy. Which counts a low value for the quality of service provided and lowers down patient satisfaction rate. f. During use of software sometimes the functioning of the system stops and if the the patient is waiting for the bill cannot be generated which increases the time of dispensing of drugs. g. The bulk storage in the side shelves includes tablets which are sometimes difficult to find. h. Difficulty in dispensing both in-patients and out-patients prescriptions at the same time. i. Availability of only two monitors.
Sagar Hospital	Most common loopholes and challenges found in the hospital pharmacy. As hospitals have a pharmacy, they mostly followed the FIP guidelines but there are some challenges and loopholes faced by the hospital pharmacy. The challenges related to the pharmacist include the inadequacy of service promotion, absence of service continuity, poor DIC service, and lack of commitment, communication, and confidence among clinical pharmacists. Most respondents declared that poor attitude toward the services, conflict of interest due to the unclear scope of practice, and absence of cooperation are the challenges that radiate from health practitioners such as nurses and physicians. Some respondents also described challenges that arise from the hospital management and its set-up. The challenges they mentioned include lack of training, shortage of skilled manpower, lack of incentives, absence of facilities in the ward for clinical

Qualitative survey result	Observations
	pharmacists, and collaboration between academics and hospital clinical pharmacists. The other challenges stated by the respondents were due to the academic policy and the curriculum itself. This encompasses some gaps of the curriculum; absence of clear job description and working guidelines; and documentation system. The major challenges for clinical pharmacy services described were a lack of support from hospital management, absence of clearly defined roles and responsibilities for the clinical pharmacists, a shortage of pharmacy workforce and staff turnover, a lack of follow-up from responsible bodies, and a lack of enough salaries and incentives for the healthcare providers.
Aditya Netralaya	a. No separate pharmacy is available. No separate pharmacist is available. b. The medicines are available at the reception only along with the billing. c. No proper records are maintained and hence it is difficult to monitor. d. The hospital is not departmentalized; hence, there is no specific manager for the pharmacy. e. Drug availability is limited. No proper records are maintained for particular drugs. So, it will take time to dispense the drugs at the counter. It increases the patient waiting time.
Mithra Hospital	1. The inventory management of medications is not properly followed the guidelines. 2. Patient counseling is not followed in most cases. 3. The quality of drug maintenance and record maintenance should be more technological and confidential. 4. The up-gradation of storage of medicines is not followed; it follows the medications of doctor's prescription only.
General	Many of the hospitals do not have an assigned clinical pharmacist, which makes the difficulty in practicing Rule number 2 of the standards of GPP. Workforce shortage during peak OPD hours there should be a training of pharmacists and associated personnel on regular basis. Updating newer guidelines. Sometimes when too much workload is there, pharmacists are unable to properly counsel the patients/attendants regarding doses, their uses, drug interactions, etc. Shortage of time for EWS pharmacy. Because medications are not always available, it is extremely inconvenient for patients to return after time. The room does not have adequate light and ventilation. There is a lack of adequate space for the storage area of the pharmacy. No trained personnel in pharmacy, that is, the staff is from nonpharmaceutical or nonmedical backgrounds. No proper support is seen in abiding by policies of government related to health outcomes as per the guidelines of Role 4. Drug availability is not meet the needs of patients. No proper record is maintained in any aspects related to dispensing or storing of drugs.
	No proper follow-up of patient turnaround time at the pharmacy
Narayana Health City	Missing of personnel in the pharmacy increases the patient waiting time
Omni R K Hospital	No trained personnel in pharmacy, that is, staff from nonpharmaceutical or nonmedical background
Gayathri Hospital	No trained personnel in pharmacy, that is, staff from non-pharmaceutical or non-medical background

Pharmacy practice, which includes clinical, community, and hospital pharmacy, is referred to as healthcare in the true sense. PPs establish a vital link between doctors, nurses, and patients in a social community group, with the goal of improving patient welfare and safety [26]. Pharmacists must provide more than only distribution; they must also provide services to society, such as patient counseling, guidance, and the organization of drug data for human services, suppliers, and patients [27]. For example, before the product patent was implemented in India in the early 2000s, the pharma industry thrived successfully after the amendment of new monetary reforms in India in the mid-1990s until the early 2000s [28]. In terms of educational outcomes, the degree of student achievement in pharmacy has improved from 2012 to 2020 and is predicted to continue to climb due to its scope. Many pharmacy graduates have expressed an interest in medication discovery and innovation through research [29]. It has been suggested that pharmacies play an important role in strengthening social security because they work around the clock to maintain a chain of elegant integrity on drugs, particularly during the COVID-19 pandemic. From the standpoint of regulatory authorities, PPs play a variety of roles and duties in the execution of new recommendations based on situational faults [30].

8.1 The practice of pharmacy

It entails the implementation, evaluation, and interpretation of pharmacy orders as well as the dispensing of prescriptions. Pharmacists are responsible for providing patient counseling, legislation, and social services to promote pharmaceutical treatment as primary care in all areas of patient health [31]. They are also in charge of segregating and labeling pharmaceuticals and medical devices, as well as guaranteeing the safety of medication and device storage and maintaining adequate records [32].

According to a Supreme Court decision on March 5, 2020, [IN THE SUPREME COURT OF INDIA CIVIL ORIGINAL/APPELLATE JURISDICTION TRANSFERRED CASE (CIVIL) NOS ... ...OF 2020 [TRANSFER PETITIONS (CIVIL) NOS. 87-101 OF 2014] The Pharmacy Council of India. Petitioner Versus Dr. S.K. Toshniwal Educational Trusts Vidarbha Institute of Pharmacy and Ors. Etc. .. Respondents]

- The Pharmacy Act is a nonconstitutional statute.
- There cannot be two regulators at the same time.
- The AICTE Act is a broad statute.
- The Pharmacy Act is a unique piece of legislation.
- The Pharmacy Act is a comprehensive code that includes pharmacy specialists.
- Only the Pharmacy Council of India (PCI) has the authority to set pharmacist qualifications.
- Only PCI has the authority to set standards for pharmacies.
- PCI will be solely responsible for approval and admission to institutes.
- The Pharmacy Act has the power to govern not just pharmacy education but also the profession [33].

DOI: http://dx.doi.org/10.5772/intechopen.100635

In India, there are two types of pharmacists—those who work in hospitals and those who work in

- Pharmaceutical industry
- Pharmacies in the community
- Pharmacovigilance is a term that refers to the monitoring of pharmaceuticals.
- Pharmacy schooling
- Pharmacovigilance procedures

8.2 India's first pharmaceutical company

Bengal Chemical and Pharmaceuticals Ltd. was founded in 1892 as an individual initiative by the "Father of Indian Chemistry," Acharya Prafulla Chandra Ray, and is considered the country's first pharmaceutical enterprise. For the first time in more than six decades, Bengal chemicals and medicines Ltd. made a profit in 2016–17 [34].

8.3 Regulation of pharmaceuticals

For the manufacture, distribution, and sale of medications, India has the harshest rules and regulations in the world. The ability of law enforcement is not just determined by good legislation; it also requires stringent regulations and the acceptance of successful challenges. According to the Mashelkar committee's recommendations, one drug inspector should be assigned to every 50 manufacturing units and one to every 200 distribution outlets. In most states, 30–50% of positions are vacant, and the number of positions available is insufficient to expand the pharmaceutical sector. According to the Mashelkar committee report, one drug inspector is assigned to every 200 retailers, and the data from the All-India Organization of Chemists and Druggists (AIOCD) indicates that there are approximately 8.5 lakh pharmacies to be considered, implying that more than 4250 drug inspectors are required to manage the retail pharmacy segment [35].

CDSCO and SDRA have different roles.

Pre-manufacturing is the process of preparing a product for manufacture.

Expert panels help the Central Drugs Standard Control Organization (CDSCO) to regulate premanufacturing. The following is their primary role in dealing with the two major criteria:

- Clinical studies (clinical trial registry, GCP, inspections, ethics committee, and significant adverse events)
- Import and marketing license approvals for new drugs [36].

8.4 Industrial production

The State Drug Regulatory Authority (SDRA) oversees regulating drug product manufacturing in the states, while CDSCO oversees union territories. In some circumstances, the CDSCO and SDRAs undertake joint inspections [37]. The terms listed below are required for the manufacture of pharmaceutical products.

- Permit (generic and marketing license)
- GMP (good manufacturing practices) inspection
- Drug testing and noncompliance penalties [31].

8.5 Sales and distribution

The SDRAs oversee regulating the distribution and sale of drugs in the states, while the CDSCO oversees the union territories. The terms listed below are required for the distribution and sale of pharmaceutical items.

- Permit (sale and distribution practices)
- Good Distribution Practices Inspection (GDP)
- Drug testing and noncompliance penalties [38].

8.6 Situation now

The current state of pharmacy education, authority functions, and future difficulties of pharmacists in all aspects was discussed, particularly considering the COVID-19 epidemic. During the development phase, it must overcome numerous obstacles, including regulations, duration, process controls, legal stumbling blocks, and situational flaws [39]. **Table 3** summarized the educational updates.

8.7 The government's initiative

From June 2, 2020, the PLI Scheme will be open for 4 months, allowing investors to propose the construction of Greenfield facilities for any of the 53 important drug intermediates and bulk pharmaceuticals that are now barely made in India [40]. The list contains important chemicals found in regularly given medications such as paracetamol, aspirin, metformin, atorvastatin, and others. The whole planned incentive package is worth Rs. 13,760 Crore, which is divided between bulk medicines and medical devices as shown in **Figure 1**.

COVID-19 PPs

- Millions of people around the world are affected
- Insight into no cure
- Online (digital) education

Pharmacy program	Number of institutions	Annual intake
D Pharm	3022	180,770
B Pharm	1961	125,524
Pharm D	267	8010
M Pharm	792	24,465
Ph. D. in Pharmacy	31	1240

Table 3.
Statistics of pharmacy education program in India.

- Evaluate Indian strategy and tactics.
- India's pharmaceutical industry's expectations in terms of global visibility
- Opportunities with challenges

8.8 Regulatory hurdles

Across the several layers of regulatory and enforcement procedures, this industry suffers from knowledge asymmetry. Because each state has its own regulations, there is a lot of variation in the quality of drug regulation across the country [41]. India's drug inspectors, who are crucial players in drug regulation, should be highly qualified and adequately compensated to reflect the dignity that comes with such a large duty. The agency has a budget limit when it comes to implementing innovative methods of regulatory monitoring [39].

8.9 Future challenges for the pharmaceutical industry

There is a great deal of anticipation, price pressure, and caution in the development of vaccines, particularly (Covaxin, Covishield, or Sputnik) for the treatment of COVID-19, which is still in Phase I and II clinical trials in India. There is a potential that poor data will emerge in a Phase III study, as several countries have done in the past, leading to vaccine development and outcome failure [42]. Nowadays, one of the most pressing issues in the management of process controls during the formulation of pharmaceuticals is the formation of impurities in the form of genotoxic and carcinogenic impurities, which are extremely toxic and harmful to patients taking medication for diabetes, gastric ulcers, and psychosis [43]. There is also a compensation challenge in India due to the loss of fixed-dose combination (FDC) drugs, as many of the 344 drugs in this category were banned and withdrawn because they were therapeutically irrational and caused toxicity in patients when used to treat chronic diseases like tuberculosis and HIV. As a result, there are still hopes of bringing FDCs to market for patient benefit with assurances of safety and efficacy [44]. If the financial expenditure is not increased, the quality of research, drug discovery, and development will suffer.

8.10 Online education

According to Holon IQ global education intelligence, the current $6 trillion education market will grow to $10 trillion by 2030. Digital platforms account for only 3% of worldwide education spending, and by 2025, they will account for $325 billion, or less than 5% of total spending [45, 46]. Although this will go a long way toward solving the country's educational challenges, the government's present worries include guaranteeing smooth delivery of education on the post-Covid road of digital education [47]. The UGC is now working on a game-changing regulation that will allow universities with a high "NAAC" score in the top 100 NIRF rankings to offer online programs. This will be a game-changer since educational jurisdiction will become worldwide, and only equipped universities will survive [48].

8.11 Observations and recommendations

The following recommendations were implemented for the PPs based on the unique circumstance faced by each country because of COVID-19:

- The pharmaceutical business must shift its focus away from COVID-19 and onto the needs of the country, as well as ideas for overcoming governmental and regulatory obstacles.
- The government should provide adequate funds to PPs researchers for the development of novel drugs and the conduct of clinical trials.
- To achieve the required results, the pharmaceutical business should learn to invest in quality staff and move its attention from the product to the patient [49].
- Application of state and federal norms to protect public health and disease prevention.
- Strengthening the regulations governing healthcare administration and professionals, requiring them to adhere to severe state-based rules.
- There should not be a large reliance on API procurement from China; instead, actions should be done to ensure API supply from within the country.
- Enabling workflow changes in principles while dealing with emergencies
- To raise awareness and develop educational training programs to protect people's physical and mental health during the pandemic.
- To combat COVID-19 as a naturopathy-based treatment, the pharmacognosy and phytopharmacy department should focus on herbal and ayurvedic formulation development.
- Positive, credible, and evidence-based marketing to be made by PPs on the conventional way of using AYUSH-based drugs to combat COVID-19.
- The general public should consume "Kapasura kudineer," which is widely available in retail pharmacies, particularly in south India, and contains many combinations of herbs that can combat the coronavirus and other microorganisms. This has shown promising results, as many people have been benefited and relieved from COVID-19 in asymptomatic reported positive cases.
- Steps to be done for e-learning/webinar programs in order to receive additional updates from the WHO and the health advisory council for pharmaceutical development and public health protection.
- In the event of an emergency, the use of PPs as second-line healers or physicians to diagnose the patient's ailment [50].
- During the COVID-19 epidemic, enabling e-prescriptions and home delivery of medicines to self-isolated individuals and older people by guaranteeing preventive steps and safety.
- Advocacy with policymakers and stakeholders must be strengthened in order to create the optimal regulatory framework for the pharmaceutical industry to thrive.

9. Unethical practices in healthcare and pharmaceuticals during COVID 19 in India

The devastating second wave of the COVID-19 outbreak has brought strangers together offline and online to assist one another in any way they can. Even terrible stories of death and loss have not deterred some people from engaging in unethical and immoral actions including black-marketing medical life-saving medicines, hoarding, and defrauding in the name of COVID-19 medical supplies, and charging exorbitant prices for ambulance service. The Delhi Police recently raided a fine-dining restaurant in Lodhi Colony and seized 419 oxygen concentrators. These were being sold at an astronomical price of 69,999, complete with bogus MRP labels. About 96 oxygen concentrators were seized from Khan Chacha, owner of a famous food joint in Khan Market. When Delhi-NCR is in the midst of the biggest crisis in its history, notable figures in the city urge tough punishments for anyone who engages in such unlawful and inhumane activities. In connection with a Rs 450 crore medical equipment maintenance scam, Andhra Pradesh's Crime Investigation Department (CID) has filed a complaint against Bengaluru-based TBS India Telematic & Biomedical Services Private Limited, private individuals, and health department employees. The case was filed under the sections of the Indian Penal Code dealing with cheating, criminal conspiracy, and criminal breach of trust. Price increases, inflated bills for nonexisting equipment, and increased invoices for equipment in government hospitals in AP are among the irregularities. Over 300 doses of anti-COVID vaccination were taken from a hospital in Jaipur during a spike in coronavirus illness, forcing authorities to file a criminal complaint. According to agency reports citing a Centre RTI response, Tamil Nadu recorded 12.10% vaccine wastage, followed by Haryana (9.74%), Punjab (8.12%), Manipur (7.80%), and Telangana (7.80%) (7.55%). Vaccine wastage in Tamil Nadu is up to 13% for Covaxin and 9% for Covishield, according to data from the state directorate of public health, which distributes vaccines and regulates its use. Between January 16 and April 17, 7.14 lakh doses of Covaxin and 44.80 lakh doses of Covishield were sent to drug retailers, according to state authorities. A vial must be used within 4 h of being opened and stored at a temperature between 2 and 8°C. The state has used 47.05 lakh dosages by April 17, including 40.5 lakh Covishield. Health staff was under pressure to vaccinate as many individuals as possible. Even though there were not enough individuals, they opened bottles of 20 and 10 dosages. We will now be able to expand immunizations due to an increase in cases and stock. On Friday, as a new day dawned over Delhi, a somber narrative surfaced from Sir Ganga Ram Hospital, one of the capital's most prestigious institutions. In the last 24 h, at least 25 COVID-19 patients had perished there. Physical ventilation—occasional6manual compression of a gas-filled reservoir bag to drive the gas into a patient's lungs—has been used on occasion to ensure that patients on ventilator assistance survive. The burning question is how to treat patients when key supplies are not available. As per specialist doctors without oxygen, a patient who requires ventilator assistance can die in minutes. Many other hospitals were in a similar situation. The hospitals then devised a plan to have two patients share a cylinder by using connectors. But also, according to doctors, those on a ventilator can die in minutes if their oxygen supply is cut off. Doctors face a significant hurdle because of this. Despite government pledges and rules, the black market for Remdesivir, a crucial COVID-19 medicine, is thriving in Chennai, with many pharma dealers asking up to Rs 14,000 per 100-mg vial. Even though the Tamil Nadu government has promised to assist them in obtaining the drug, many private hospitals have thrown in the towel, leaving numerous COVID-19 patients in their care without treatment. A private COVID panel hospital has been charged with overbilling by the

Kanpur district administration. On Monday, the complainant filed a complaint alleging that he was overcharged by Rs 3.50 lakh in the previous 7 days, and on Tuesday, the district administration issued a notice to the hospital's management after completing a preliminary investigation. The charges were confirmed to be accurate after a preliminary investigation by the extra city magistrate and a government doctor. When the hospital's management failed to respond satisfactorily, district magistrate Alok Tiwari issued an order to register an FIR against it on Tuesday. The administration has appointed static magistrates and sector magistrates to investigate overbilling issues in private panel hospitals, according to the district magistrate. Furthermore, the government has already circulated cell phone numbers for the public to use to report any overbilling complaints. Despite threats of severe consequences for overbilling, private panel hospitals are seizing every opportunity to overcharge patients. Prices of oxygen concentrators, oximeters, and nebulizers have risen by 50–100% in the last 10 days, owing to a massive demand-supply mismatch as the coronavirus epidemic enters its second wave. Even as the organized medical device sector, legal entities, and consumers pressed for enforcing pricing control over these products, e-commerce giant Amazon stated it has begun removing listings of accounts selling these products beyond MRP. Over the last 7 days, the price of oxygen concentrators, which generate oxygen from the air, has nearly doubled, while the price of oximeters has increased by Rs 1000–2000. Prices have risen in both physical businesses and online marketplaces such as Amazon and Flipkart. The government set a price cap on certain products last year, but some vendors, companies, and importers are not following it. A spokesman for Amazon India stated the business is taking urgent action to stop the surge pricing. Hundreds of tweets from customers have gone viral, claiming that they had to pay more than Rs 1 lakh for oxygen concentrators that would normally cost Rs 45,000. The device's monthly rentals have also increased from Rs 5000 to Rs 10,000–20,000. Many sellers and importers have boosted costs of COVID-essential medical supplies by two to four times in just a week, even on online markets like Amazon and Flipkart. The problem is large with imported products by opportunistic dealers who import these products at low costs and then continuously raise prices to profit from the scenario while blaming it on their suppliers.

10. Essential commodities (control of unethical practices in marketing of drugs), 2017

Since the pharmaceutical industry's exponential rise, there has been a need to regulate the interaction between pharmaceutical firms and medical practitioners. The Indian Medical Council (Professional Conduct, Etiquette, and Ethics) Regulations, 2002, as revised until February 1, 2016, ("MCI Regulations") were enacted with this in mind. The MCI Regulations, Regulation 6 (Unethical Acts), establishes a code of conduct for medical practitioners in their interactions with the pharmaceutical and allied healthcare industries. The MCI Regulations, on the other hand, apply only to medical practitioners. As a result, on December 12, 2014, the Department of Pharmaceuticals ("DoP") declared the Uniform Code of Pharmaceuticals Marketing Practices ("UCPMP") effective on January 1, 2015, for a 6-month period of voluntary adoption. When the DoP announced UCPMP, it stated that it was a voluntary code for the Indian pharmaceutical industry, but warned that if it is discovered that the pharma associations/companies have not implemented it successfully, the government may consider making it a statutory code. The DoP then extended UCPMP until further orders on August 30, 2016. Despite the fact that the UCPMP was the first of its kind and contained substantial regulations, the DoP attempted to replace the voluntary code with mandatory recommendations since it was optional. As a result, the DoP transmitted the draught Essential Commodities (Control of Unethical Practices in Drug Marketing) Order, 2017 ("Order") to the Law Ministry in July 2017 for final approval. The Law Ministry, on the other hand, is concerned about the Order, which is now being sought to be issued under the Essential Commodities Act, 1955. It is worth noting that the Order is beyond the scope of the Vital Commodities Act of 1955 ("Act"), whose goals are to control the manufacture, supply, and distribution of essential commodities rather than to oversee medication marketing. In 2015, the UCPMP was expanded to include the medical devices business, which is worth more than Rs 25,000 crore annually. The Order indicates that it will not apply to medical devices because the DoP is in the process of creating and releasing a separate code of marketing practices for the medical device industry (which would be voluntary for 6 months). The Order, like the UCPMP, prohibits pharmaceutical corporations from presenting cash, gifts, or sponsorship to doctors, chemists, or pharmacists, as well as giving travel facilities or paid vacations. Any pharmaceutical firm or its representatives are prohibited from giving presents, cash cards, hampers, or any other item that may produce monetary benefit or allow profits in kind to a medical practitioner, a retail chemist, or pharmacists, or their "family members" under paragraph 3 of the Order. It does, however, allow for the financing of academic conferences organized by medical groups, as well as the organization of screening camps or awareness campaigns in government-owned healthcare institutions, with the caveat that these cannot be used as surrogate advertising. A pharmaceutical business or its agent is also prohibited from providing free samples to any medical practitioner under the terms of the Order. Allowing pharmaceutical companies to distribute free samples up to a full course of medicines for a maximum of three patients has created an exemption to this norm. Three, on the other hand, is a modest amount for any medical professional to comprehend and analyze the effects of a new drug on patients. It is important to note that the term "Agent" is defined very broadly in paragraph 2(b) of the Order, and includes any person, company, society, nongovernmental organization, or other institution who has been hired or authorized by a third party to call on any healthcare facility to promote a pharmaceutical company's drugs. The Order further states that a company's MD or CEO is accountable for ensuring that the Order is followed. All complaints involving violations of the Order shall be investigated by an Ethics Compliance Officer (ECO) designated by the Government of India, who shall not be below the rank of a Joint Secretary to the Government of India, according to paragraph 5 of the Order. It should be noted that a firm or its agent who fails to comply with the terms of subparagraphs (a) (b) (c) or (d) of paragraph 3 of the Order will face a penalty under paragraph 4 of the Order. Furthermore, paragraph 5(4) of the Order lays out the mechanism for imposing a penalty by prohibiting the breaching corporation from marketing its best-selling product for the previous 12 months for a period of three to one year. The Order also states that corporations can pay a penalty ranging from Rs 5 lakh to Rs 10 lakh to have the marketing prohibition lifted. According to the Order, the DoP Secretary will be the appellate authority, and appeals from the Appellate authority's orders will be handled by the court.

11. The way forward

There is an utmost necessity to relook and re-establish the standards of pharmacy practice in healthcare setups available in each and every corner of the country in line with guidelines provided by the World Health Organization (WHO) and

International Pharmaceutical Federation (FIP). For that, the dependency and responsibilities are very high on healthcare professionals, particularly in this pandemic situation. The pharmacy zone is adaptable, evolving, and increasingly diverse, offering a wide range of work and management opportunities to execute [27]. PPs are human service professionals whose responsibilities include safeguarding individuals by dispensing medications based on prescriptions [28]. Representing the world's third-largest medicinal services with active gathering, and in India, there are over 1,000,000 (1 million) enrolled PPs employed in various capacities and readily contributing to the country's well-being [51]. Pharmacy practice, which includes clinical, community, and hospital pharmacy, is referred to as total healthcare in its true sense. Through adaptation and implementation of GPP in healthcare setup, PPs form an essential link between physicians, nurses, and patients in the social community group, with an ultimate emphasis on patient well-being and protection [30].

There should be country-wise GPP training programs are specifically aimed to gain rapid insights from World Health Organization (WHO) and International Pharmaceutical Federation (FIP) joint guidelines for advancing pharmacy practice worldwide and especially in India. In the view of current COVID-19 crisis, this initiative should be the first of its kind training program being offered in India. This training program mainly focuses on experiential training while promoting the GPP concept among healthcare professionals in India. This course will help the participants to get easy access to the guidelines, provides opportunities for peer discussions, and eventually be influential in improving pharmacy practices.

There should be a very focused approach for the same as follows

1. To understand the roles of pharmacists, national pharmaceutical organizations, and healthcare systems in developing GPP standards.

2. To understand the role of the pharmacist as a part of healthcare and the requirements of this role.

3. To understand the methodology and principles of quality management.

4. Be able to set standards, measure the quality and use the principle of continual improvement in one's own working environment.

5. This course mainly explores and discusses the WHO/FIP GPP guidelines, the model of GPP in the hospital setting, and aseptic preparations.

6. Additionally, focus on WHO, MoH & FW, ICMR & NCDC guidelines for handling, treatment, and disposal of biomedical waste at healthcare facilities.

Across the globe, pharmacy is one of the most important, dynamic, and versatile segments of the healthcare industry. The pharmacy zone has become adaptable, evolving, increasingly diverse, adhering to the quality standards. However, the importance of ensuring appropriate quality to every patient has taken a center stage in this pandemic, especially in India. In such time, it is important to understand and execute the methodology and principles of quality management and set standards, for continual improvement in one's own working environment, while realizing the emerging roles of pharmacists, national pharmaceutical organizations, and healthcare systems in developing good pharmacy practice (GPP) standards in this pandemic era.

The International Pharmaceutical Federation and World Health Organization define good pharmacy practice (GPP) as practices that meet the personal needs of patients or those using pharmacy services by offering appropriate evidence-based care. In developed countries, pharmaceutical assistance is defined as a pharmaceutical practice model that involves attitudes, ethical values, behaviors, skills, appointments, and co-responsibility to prevent diseases, promote and recovery health in an integrated manner as part of the healthcare process, highlighting, among other, the requirement that the institution fully adopts the GPP. The program should be designed to take care of the Indian healthcare system and its context of adoption of "new normal" due to the unprecedented event of COVID 19 and the importance of GPP for the healthcare professionals in same.

This kind of training program should be opened to

- Health, Health IT, Hospital & Pharma Management Students & Professionals.
- Medical, Dental, Physiotherapy, Nursing, Pharma, Paramedical Students.
- Any Graduate student/Professional who wants to pursue a career in healthcare.

Today, the forms of care are shifting from secondary care providers to primary care providers to patients (**Figure 2**). This trend has already started in developed countries, such as the behind-the-counter drug option in European countries that was already endorsed by the FDA. The FDA showed positive signs toward boosting the numbers of nonprescription drug statuses to over-the-counter statuses. The healthcare delivery system is coming closer to the patient due to knowledge and understanding, as well as better diagnostic tools and monitoring devices [52] (**Figure 2**).

The use of simulation and related technology in healthcare education will continue to grow in the next years, and this methodology has a collective role within the pharmacy curriculum. It is expected that increasing the quantity of simulation in pharmacy curricula will have a good impact on pharmacy student education and training, resulting in favorable outcomes for patients and the healthcare team. The obvious goal of incorporating simulation techniques into the pharmacy student training curriculum is to increase pharmacist education and training with the ultimate goal of enhancing patient care and safety. Simulation experiences will never be able to replace real-world clinical experiences, but they do have the ability to

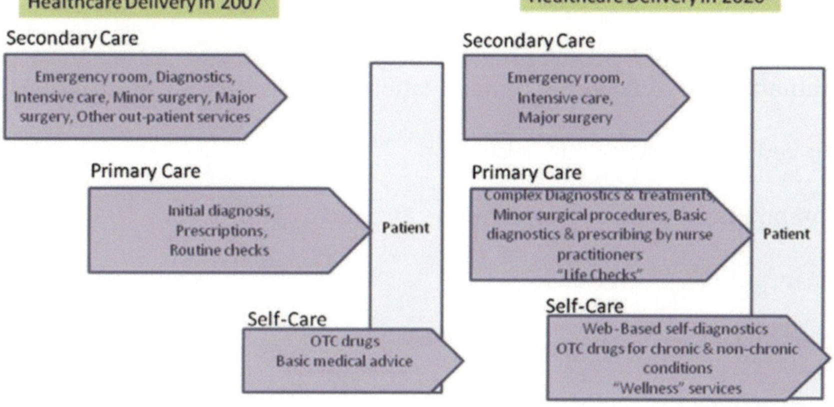

Figure 2.
The healthcare delivery status of 2007 and 2020. Source: PricewaterhouseCoopers.

supplement clinical education and serve as a tool for developing the skills necessary for a successful pharmacist. Simulation approaches have been utilized in pharmacy education to address broad cognitive and social skills, particularly communication, decision-making, ethical dilemmas, prioritization, and teamwork, in addition to the development of technical skills such as procedural and clinical abilities. Simulated learning environments could provide a more systematic approach to both clinical skill training and pharmacy programs that aim to provide an opportunity for theoretical knowledge to be applied to a real clinical situation. Basic sciences, dispensing, and drug supply all benefit from simulation's constant, predictable experience. In an ideal world, simulation training would be integrated into all levels of pharmacy education and training.

Types of simulation technology used in healthcare education [53, 54].

1. High-fidelity patient simulator or mannequin—able to mimic human actions and physiology and respond to physiologic and pharmacologic interventions
2. Task trainer —designed to help learners practice, specific skills and do not have the extensive programming capabilities of high-fidelity models. It can be considered as *low-fidelity simulators or moderate-fidelity* simulators depending on the sophistication of the model
3. Standardized patients—live people who are coached to portray patients, usually referred to as simulated patients
4. Virtual reality simulator—in which a computer display simulates the physical world and user interactions are with the computer within that simulated (virtual) world
5. Full environment simulation—it involves the incorporation of high-fidelity mannequins, standardized patients, healthcare professionals, and ancillary equipment to recreate a real-life clinical environment

11.1 Key components should be included in hospital & community pharmacy practice: automation & technology

- Serves to increase efficiency and accuracy of dispensing
- Re-direct staff time away from routine technical tasks and toward more direct patient care activities
- Featured Systems: ADC, place medications much closer to the user, but still allow electronic verification
- Pharmacy Robot: Reduced preparation and check time for medications, minimizations of potential contamination in sterile product preparation
- Bar coded medication administration
- Computerized prescriber order entry systems (CPOE)
- Smart pumps
- Clinical decision support system (CDSS)
- Predictive population risk stratification
- Patient self-management tools

list of contributors in who gpp 2011/ippr 2015 regulation adherence survey	
Mr. B. Hajee Bahadoor	**Dr. Radhika Sanjay Hatolkar**
BSc., PGDBM Health & Hospital Management, IIHMR Bangalore	*Bachelor of Physiotherapy, PGDBM Health & Hospital Management, IIHMR Bangalore*
Dr. Chandolu Teja Venkata Narayana	**Ms. Ranjitha Roy**
Bachelor of Physiotherapy, PGDBM Health & Hospital Management, IIHMR Bangalore	*Bachelor of Nursing, PGDBM Health & Hospital Management, IIHMR Bangalore*
Dr. Chandwani Mayuri Mukeshkumar	**Mr. S. Neeraj Naga Santosh**
Bachelor of Dental Surgery, PGDBM Health & Hospital Management, IIHMR Bangalore	*B.Tech, PGDBM Health & Hospital Management, IIHMR Bangalore*
Dr. Geetanjali Bhaskar	**Dr. S. Siri Charan**
Bachelor of Dental Surgery, PGDBM Health & Hospital Management, IIHMR Bangalore	*Bachelor of Dental Surgery, PGDBM Health & Hospital Management, IIHMR Bangalore*
Ms. Goyal Anshita Jitendra	**Dr. Sanem Pavani Mayusha**
BCom., PGDBM Health & Hospital Management, IIHMR Bangalore	*Bachelor of Homoeopathic Medicine & Surgery, PGDBM Health & Hospital Management, IIHMR Bangalore*
Ms. Iffat Naz	**Dr. Saniya Suleman Ahmed**
BCom., PGDBM Health & Hospital Management, IIHMR Bangalore	*Bachelor of Dental Surgery, PGDBM Health & Hospital Management, IIHMR Bangalore*
Dr. Kanakam Viswa Praneeth	**Mr. Seelambuttagalla Bharath**
Bachelor of Dental Surgery, PGDBM Health & Hospital Management, IIHMR Bangalore	*BA, PGDBM Health & Hospital Management, IIHMR Bangalore*
Ms. Madhushree	**Dr. Shreya Shresth**
Bachelor of Hospital Management, PGDBM Health & Hospital Management, IIHMR Bangalore	*Bachelor of Medicine & Bachelor of Surgery, PGDBM Health & Hospital Management, IIHMR Bangalore*
Dr. Mansi Trakroo	**Mr. Sidharth Samantaray**
Bachelor of Dental Surgery, PGDBM Health & Hospital Management, IIHMR Bangalore	*B.Tech Computer Science & Engineering, PGDBM Health & Hospital Management, IIHMR Bangalore*
Ms. Mridula Pandey	**Ms. Sweta Kumari**
BSc., PGDBM Health & Hospital Management, IIHMR Bangalore	*BBA, PGDBM Health & Hospital Management, IIHMR Bangalore*
Dr. Apoorva Uday Munagekar	**Dr. Sweta Sharma**
Bachelor of Physiotherapy, PGDBM Health & Hospital Management, IIHMR Bangalore	*Bachelor of Medicine & Bachelor of Surgery, PGDBM Health & Hospital Management, IIHMR Bangalore*
Mr. Nilesh Sharma	**Ms. Anjali Deb**
BSc. Biotechnology, PGDBM Health & Hospital Management, IIHMR Bangalore	*BA, PGDBM Health & Hospital Management, IIHMR Bangalore*
Dr. Nunna Krishna Madhuvani	**Dr. Prajwal Ramesh**
Bachelor of Physiotherapy, PGDBM Health & Hospital Management, IIHMR Bangalore	*Bachelor of Dental Surgery, PGDBM Health & Hospital Management, IIHMR Bangalore*
Ms. Pallavi Kushwaha	**Dr. Shaila Sharma**
BBA, PGDBM Health & Hospital Management, IIHMR Bangalore	*Bachelor of Dental Surgery, PGDBM Health & Hospital Management, IIHMR Bangalore*
Dr. Parul Phulera	**Dr. Shradha Saran**
Bachelor of Dental Surgery, PGDBM Health & Hospital Management, IIHMR Bangalore	*Bachelor of Dental Surgery, PGDBM Health & Hospital Management, IIHMR Bangalore*

list of contributors in who gpp 2011/ippr 2015 regulation adherence survey	
Dr. Poosarla Sree Harsha	**Brig. Sandeep B Singh**
Bachelor of Dental Surgery, PGDBM Health & Hospital Management, IIHMR Bangalore	*B.Tech Electronics, PGDBM Health & Hospital Management, IIHMR Bangalore*
Ms. Preetha S.	**Ms. Radha Virendra Deosthalee**
BSc. Perfusion Technology, PGDBM Health & Hospital Management, IIHMR Bangalore	*Bachelor of Hospital Management, PGDBM Health & Hospital Management, IIHMR Bangalore*
Ms. Priyanka Saha	**Dr. Ramesetty Lavanya**
B.Tech Civil Engineering, PGDBM Health & Hospital Management, IIHMR Bangalore	*Bachelor of Dental Surgery, PGDBM Health & Hospital Management, IIHMR Bangalore*
Lt. Col. Eswar Chandar	**Mr. Peeyush Srivastava**
B.Sc. (National Defence Academy), PGDBM Health & Hospital Management, IIHMR Bangalore	*BSc., PGDBM Health & Hospital Management, IIHMR Bangalore*
Ms. Shipra Mishra	**Ms. Kesari Sai Sowjanya**
Bachelor of Hospital Management, PGDBM Health & Hospital Management, IIHMR Bangalore	*BSc. Clinical Nutrition & Dietetics, PGDBM Health & Hospital Management, IIHMR Bangalore*
Mr. Yohan Deepak	
Bachelor of Hospital Administration, PGDBM Health & Hospital Management, IIHMR Bangalore	

Author details

Mrinmoy Roy
Indian Institute of Health Management and Research, Bangalore, India

*Address all correspondence to: mroy2612@gmail.com

References

[1] Death of Covid patients due to non-supply of oxygen not less than genocide: Allahabad HC—Coronavirus Outbreak News (indiatoday.in)

[2] The National Portal of India, New Delhi. Know India-states and Union Territories. 2009. Available from: http://india.gov.in/knowindia/state_uts.php [Accessed: November 01, 2014]

[3] Visaria P. Demographics of ageing in 12 India. Economic and Political Weekly. 2001;**36**(22):1967-1975

[4] Rana RD. Pharmacy in ancient India. Indian Journal of History of Science. 1987;**22**(2):119-121

[5] Saha D, Paul S. Glimpse of pharmaceutical education in India: History to advances. International Journal of Pharmacy Teaching and Practices. 2012;**3**(4):387-404

[6] Report of the Steering Committee on Health for the 12th Five Year Plan. Yojana Bhavan, New Delhi: Health Division, Planning Commission; 2012. p. 7. Available from: http://planningcommission.nic.in/aboutus/committee/strgrp12/str_health0203.pdfc

[7] Health Infrastructure, National Health Profile. 2011. Available from: http://cbhidghs.nic.in/writereaddata/mainlinkFil e/11%20%20Health%20Inf rastructure%202011.pdf [Retrieved: February 20, 2015].

[8] Sreelalitha N, Vigneshwaran E, Reddy GNYP, Reddy MR. Review of pharmaceutical care services provided by pharmacists. IRJP. 2012;**3**(4):78-79

[9] Nand P, Khar RK. Hospital and Clinical Pharmacy. 5th ed. Delhi: Birla Publications Pvt Ltd; 2008

[10] Nahata MC. Hospital pharmacy practice in India. Drug Intelligence & Clinical Pharmacy. 1984;**18**(6):523-524

[11] Annual Report 2011–12, Department of Pharmaceuticals, Ministry of Chemicals and Fertilizers. Government of India. http://pharmaceuticals.gov.in/annualre port2012.pdf [Accessed: March 1, 2015]

[12] Desale P. An overview about pharmacy education in India. Indian Journal of Research in Pharmacy and Biotechnology. 2013;**1**(3):329-330

[13] Sachan A, Sachan AK, Gangwar SS. Pharmacy education in India and its neighbouring countries. International Current Pharmaceutical Journal. 2012; **1**(9):294-301

[14] Goyal R, Pareek P. A review article on prescription behaviour of doctors, influenced by the medical representative in Rajasthan, India. IOSR Journal of Business and Management. 2013;**8**(1):56-60. Available from: http://www.iosrjourna ls.org/ ccount/click.php?id=5229 [Accessed: December 15, 2014]

[15] The Order of MOH of Ukraine of 30.05.2013 № 455. On approval of the guidelines good pharmacy practice: Quality standards pharmacy services. Weekly Pharma. 2013;**27**:18-20

[16] Vetutneva NO, editor. Modern Concept of Quality Assurance of Medicinal Products. Vinnitsa: Nilan LTD.; 2018. pp. 69-75

[17] General Pharmaceutical Council of Spain. Good Pharmacy Practice in Spanish Community Pharmacy. Madrid: General Pharmaceutical Council of Spain; 2013. p. 15

[18] Good Pharmacy Practice (GPP) in developing Countries. Recommendations for Step-Wise Implementation: FIP. 1998:16

[19] Joint FIP/WHO guidelines on good pharmacy practice: Standards for

quality of pharmacy services. Forty-fifth report of the WHO expert committee on specifications for pharmaceutical preparations. WHO Technical Report Series. 2011;**961**: 310-323

[20] The Patient in Focus: A Strategy for Pharmaceutical Sector Reform in Newly Independent States//EUR/ICP/QCPH062202WHO/DAP/98/8. 1998. p. 38

[21] Good Pharmacy Practice in the Newly Independent States. Guide the Development and Implementation of Standards. Provisor; 2002. p. 17: 3-8, 18: 3-13, 21: 8-14

[22] FIP Reference Guide on Good Pharmacy Practice in Community and Hospital Settings. First edition 2009/ Approved by FIP Council in Istanbul in September 2009. The Hague, Netherlands, International Pharmaceutical Federation Internationale Pharmaceutique; 2009. p. 62

[23] Forty-fifth Report of the WNO Expert COMMITTEE on Specifications for Pharmaceutical Preparations (WHO Technical Report Series; No 961). 2011. pp. 310-23

[24] Joint FIP/WHO Guidelines on Good Pharmacy Practice: Standards for Quality of Pharmacy Services. Working Document QAS/10.352/. Vol. 1. 2010. p. 19

[25] Joint FIP/WHO Guidelines on Good Pharmacy Practice: Standards for Quality of Pharmacy Services from the WHO Technical Report Series, No. 961, 45th Report of the WHO Expert Committee on Specifications for Pharmaceutical Preparations © World Health Organization; 2011. p. 20

[26] Kasthuri A. Challenges to healthcare in India-the five A's. Indian Journal of Community Medicine. 2018;**43**:141-143

[27] Basak SC, Sathyanarayan D. Community pharmacy practice in India: Past, present and future. Southern Med Review. 2009;**2**(1):11

[28] Rotta I, Salgado TM. Ensuring consistent reporting of clinical pharmacy services to enhance reproducibility in practice: An improved version of DEPICT. Journal of Evaluation in Clinical Practice. 2015;**21**: 584-590

[29] Mohanan M, Hay K, Mor N. Quality of health care in India: Challenges, priorities, and the road ahead. Health Affairs. 2016;**35**:1753-1758

[30] MoHFW. Home; 2020. Available from: https://www.mohfw.gov.in/ [Accessed: August 05, 2020]

[31] Lakshminarayanan S. Role of government in public health: Current scenario in India and future scope. Journal of Family and Community Medicine. 2016;**18**:26-30

[32] Kalaiselvan V, Srivastava S, Singh A. Pharmacovigilance in India: Present scenario and future challenges. Drug Safety. 2019;**42**:339-346

[33] Palsokar G, Tajne M. Postgraduate research in pharmaceutical sciences in India: A survey of select pharmacy professionals. International Journal of Current Pharmaceutical Research. 2017; **9**:161-164

[34] Jose J, Rafeek N. R pharmacovigilance in India in comparison with the USA and European Union: Challenges and perspectives. Therapeutic Innovation & Regulatory Science. 2019;**53**:781-786

[35] Supreme court of India, Union of India vs. M. V. Mohanan Nair. 2020. Available from: https://indiankanoon.org/doc/55062525/ [Accessed: August 05, 2020]

[36] Kishore J. Legislation and health promotion in India. Review of Global Medicine and Healthcare Research. 2012;**37**:75-87

[37] Peraman R, Dugga HHT. Need for the refinement of licensing procedure for pharmacists in India. Journal of Young Pharmacists. 2016;**8**:293

[38] Roderick P, Mahajan R. India should introduce a new drugs act. Lancet. 2014; **383**:203-206

[39] Bennett D, Yin W. The market for high-quality medicine: Retail chain entry and drug quality in India. Review of Economics and Statistics. 2019;**101**:76-90

[40] Thaker Z, Jethva K, Bhatt D, Zaveri M, Deshpande S. Orphan drugs: Overview and regulatory review process. International Journal of Pharmaceutical Sciences and Research. 2019;**10**:505-518

[41] Praveen A, Deepak B, Abdul R. Drug approval procedure in India (CDSCO). Research & Reviews: A Journal of Drug Design & Discovery. 2015;**2**:20-26

[42] Pravat KJ. Impact of pandemic COVID-19 on education in India. International Journal of Current Research. 2020;**12**:12582-12586

[43] World Health Organization. Coronavirus Disease (COVID-19) Outbreak Situation. 2020. Available from: https://www.who.int/emergencies/diseases/novel-coronavirus-2019 [Accessed: August 10, 2020]

[44] Hedima EW, Adeyemi MS, Ikunaiye NY. Community pharmacists: On the frontline of health service against COVID-19 in LMICs. Research in Social and Administrative Pharmacy. 2020;**19**:30381-30388

[45] Ung COL. Community pharmacist in public health emergencies: Quick to action against the coronavirus 2019-nCoV outbreak. Research in Social and Administrative Pharmacy. 2020;**16**: 583-586

[46] Mukattash TL, Jarab AS. Pharmacists' perception of their role during COVID-19: A qualitative content analysis of posts on facebook pharmacy groups in Jordan. Pharmacy Practice. 2020;**18**:1900

[47] Shrimali, Gireesh, Retna Kumar, Aravind. Battery Storage Manufacturing in India: A Strategic Perspective. 2020. Available from: https://ssrn.com/abstract=3616404 [Accessed: August 10, 2020]

[48] Usha Sharma. Pharma cos should develop strategies to develop APIs and making India self-reliant: Dr Eswara Reddy. Express Pharma; 2020. Available from: https://www. expresspharma.in/latest-updates/pharma-cos-should-develop-strategies-to-develop-apis-and-making-india-self-reliant-dr-eswara-reddy/ [Accessed: August 10, 2020]

[49] Department of Pharmaceuticals. PLI Scheme for Pharmaceuticals and Medical Devices; 2020. Available from: https://pharmaceuticals.gov.in/schemes#:~:text=Download%20(1.6%20MB)-,Production%20Linked%20Incentive%20(PLI)%20Scheme%20for%20promotion%20of%20domestic%20manufacturing,Promotion%20of%20Bulk%20Drug%20Parks [Accessed: August 10, 2020]

[50] Mandaviya M. Covid-19: Centre Chalks Out Four Schemes to Boost Production of Bulk Drugs and Medical Devices; 2020. Available from: https://economictimes.indiatimes.com/news/economy/policy/covid-19-centre-chalks-out-four-schemes-to-boost-production-of-bulk-drugs-and-medical-devices/articleshow/74867468.cms?from=mdr [Accessed: August 10, 2020]

[51] Manoharan V, Ganesh K. Pharmacy professions in India during COVID-19

pandemic: Present status, future challenges and a way forward. International Journal of Applied Pharmaceutics. 2021:32, 10.22159/ijap.2021v13i1.39688-35

[52] Pharma 2020: challenging business. 2007. Available from: http://www.pwc.com/gx/en/pharma-life-sciences/pharma-2020-business-models/index.jhtml

[53] Lin K, Travlos DV, Wadelin JW, Vlasses PH. Simulation and introductory pharmacy practice experiences. American Journal of Pharmaceutical Education. 2015;75:209

[54] Vyas D, Bray BS, Wilson MN. Use of simulation-based teaching methodologies in US colleges and schools of pharmacy. American Journal of Pharmaceutical Education. 2013;77:53